AF781514

Selective Trial Test Package

Set # 1

Sibashis Nanda
Founder & CEO, Learnikx Education

Student Name : ____________________

Learnikx Education
Tutoring for gifted and talented students
https://learnikx.com.au

Selective Trial Test Package Set 1
Learnikx Education

ISBN **978-1-922608-34-5**

Trademarks
All terms mentioned in this book that are known to be trademarks or service marks have been appropriately capitalized. Learnikx Education cannot attest to the accuracy of this information. Use of a term in this book should not be regarded as affecting the validity of any trademark or servicemark.

Warning and Disclaimer
[1] Every effort has been made to make this book as complete and as accurate as possible, but no warranty or fitness is implied. The information provided is on an "as is" basis. Learnikx Education and its owners shall have neither liability nor responsibility to any person or entity with respect to any loss or damages arising from the information contained in this book.
[2] This is not an officially endorsed training or publication of the NSW Selective Test. This book is developed independently of Australian governments or other agencies.

Contents

We Want to Hear from You!

Welcome to the first edition of *Selective Trial Test Package Set 1!* As the reader of this book, you are our most important critic and commentator. We value your opinion and want to know what we're doing right, what we could do better, what areas you'd like to see us publish in, and any other words of wisdom you're willing to pass our way. You can email and let us know what you did or didn't like about this book, as well as what we can do to make our books stronger.

Please note that we cannot help you with technical problems related to the topic of this book, and we might not be able to reply to every message.

When you write, please be sure to include this book's title and author as well as your name and contact information.

Email: admin@learnikx.com.au

About Learnikx Education

Learnikx is tutoring centre for Gifted and Talented students only. Admission to paid tutoring is based on entrance test + mandatory problem-solving interview. A minimum 70% score (80% for Y6 onwards) is required for admission to tutoring. Entrance test is a selection/elimination test, not an assessment test. Entrance test happens 4 times a year and is free. Classes after entrance test are fully online.

Overall selection process:

Entrance test is a two-step process. Both are elimination rounds.

1. Aptitude test : 70% cutoff for Y4 (for Y5 tutoring) & Y5 students; 80% cutoff for Y6, Y7 and Y8 students.
2. Problem solving interview : real-life problem solving (not Mathematics); evaluates multi-dimensional thinking.

Sibashis Nanda, Educator

Siba, as he is commonly known, has over two decades of successful startup experience. He is an edutech entrepreneur.

Current Researches : A mathematical model to study the effect of porous parameter on blood flow through an atherosclerotic arterial segment having slip velocity

An analytical study on Brain Injury due to an input rotational acceleration on skull-brain system of the Human head

Use of Some Standard Mathematical Models in Physiology and Pathology INDUSTRY4.0 AND BUSINESS ANALYTICS FOR HEALTH CARE Paper ID 07 Nova Science Publishers, Inc.

Preptive Prepare Better

Series A Reading Comprehension Trial Test

Name: ______________________________

Date: ______________________________

Direction for Questions: 1 to 5

Read the two extracts and then answer the questions.

Extract A: The Telephone

American technology has a rich history with the telephone of today. It is American because it was invented in the United States by Alexander Graham Bell. Throughout history, messengers employed the mode of foot travel. They were individuals who would dash from location to location while conveying messages to others. They ran for distances and possessed extraordinary stamina. Later, as a means of communication improvement, letters were delivered via carriage. Over time, mail would be transported via train.

The advent of a novel communication instrument enabled individuals to transmit messages with remarkable speed. That device was the telegraph, which operated along a line connecting two locations. It required years to install every telegraph line in the United States. At one end of the line, an individual transmitted a message by code using a device. The message would be received and translated by an individual at the receiving end. People were able to transmit messages virtually instantly via telegraph lines, although it was not "person to person." Alexander Graham Bell attempted to obtain a functional machine, and despite numerous failures, he remained persistent. After numerous attempts, the mechanism was perfected. Everyone in the world desired telephones.

The subsequent modifications to the telephone mirrored those to the telegraph. Depicting Chicago in the year 1900 would reveal an abundance of cables in the downtown area. The initial telephone lines connected phones from one individual to another. The term for those was landlines. They subsequently constructed switching facilities so that a single line could reach multiple locations. Nonetheless, years would pass before the phone system became more efficient.

Present-day cell phones are wire-free devices. They employ airborne signals transmitted by systems reliant on the principles of physics. Numerous individuals no longer own landlines and instead rely solely on their mobile devices. The cell phone is influencing not only how people communicate but also their way of life. It can connect individuals via voice and text. Certain individuals even utilize their devices to engage in video-based conversations with others. These are the most recent communication innovations. Thus, the telephone continues to evolve.

Extract B: Health Hazards Linked to Utilization of Mobile Phones

Cell phones, also known as mobile phones, have become an essential component of daily lives of all individuals. Over fifty percent of the population in many nations uses a mobile phone, and the market for mobile phones is expanding swiftly. Saudi Arabia has the greatest percentage of mobile users among Gulf region countries, according to a study by the United Nations Conference on Trade and Development (UNCTAD). Second among Gulf states was Oman, followed by Kuwait and the United Arab Emirates. Given the global user base of mobile phones, which amounts to billions, even a marginal escalation in the occurrence of detrimental health effects could have profound and enduring consequences for public health. In addition to the volume of daily cell phone communications, the duration of individual calls and the duration of cell phone usage are significant determinants that contribute to an increased health-related risk.

As a form of non-ionizing electromagnetic radiation, radiofrequency energy emitted by mobile phones can be absorbed by tissues in close proximity to the device. Numerous variables affect the quantity of radiofrequency energy to which a mobile phone user is exposed, including the phone's technology, the distance between the user and the phone, the nature and frequency of mobile phone use, and the user's proximity to cell phone towers. The classification of mobile phone radiation as possibly carcinogenic by the International Agency for Research on Cancer (IARC) in 2011 indicates that there is a "could be some risk" of carcinogenicity; therefore, further investigation is required regarding the long-term and excessive usage of mobile phones.

Although there is no evidence to suggest that mobile phone use increases the risk of developing brain tumors, the increasing prevalence of mobile phone usage and the dearth of data for use exceeding 15 years necessitate additional research on the relationship between mobile phone use and brain cancer risk. WHO has encouraged additional research on this demographic in light of the growing prevalence of mobile phone usage among younger individuals and the possibility of extended lifetimes of exposure; it is presently evaluating the health effects of radio frequency (RF) fields on all examined endpoints. There was no correlation between cell phone usage and the occurrence of glioma, meningioma, or acoustic neuroma, according to the analyses, not even among those who had subscribed to cell phones for at least thirteen years.

According to the American Cancer Society (ACS), the IARC classification indicates the presence of potential cancer risk; however, the evidence is insufficient to establish a causal relationship and further investigation is warranted. Those concerned about radiofrequency exposure can take steps to reduce their exposure, such as limiting cell phone use, especially among children, and using an earpiece.

Mobile phone usage has been associated with negative health effects, such as alterations in sleep patterns, brain activity, and reaction periods, as reported by scientists. Further investigations are being conducted in an effort to validate these results. There exists a potential for interference with the operation of specific medical devices, such as pacemakers, implantable defibrillators, and certain hearing aids, when mobile phones are utilized in close proximity to them. Additionally, interference between mobile phone signals and aircraft electronics is a possibility. Certain nations have authorized the use of mobile phones on aircraft while in flight by implementing systems that regulate the power output of the phones.

There is a three to fourfold increased risk of traffic accidents when mobile phones are used while driving, according to research, due to the resulting distraction. The use of cell phones may expose children to a higher risk of developing brain cancer compared to adults. Due to the fact that their nervous systems are in the process of developing, they are more susceptible to carcinogenic agents.

The FDA has recommended the following measures to minimize radio frequency energy exposure: Cell phones should only be utilized during brief conversations or in situations where a landline phone is unavailable. By employing a hands-free device, the user effectively increases the distance between the phone and their head.

Question: 1 of 30

Who is credited with inventing the telephone?

- **A.** Thomas Edison
- **B.** Alexander Graham Bell
- **C.** John Logie Baird
- **D.** Samuel Morse

Question: 2 of 30

How has the telephone impacted communication today?

- **A.** It has led to a decrease in face-to-face communication.
- **B.** It has made communication less convenient.
- **C.** It has led to an increase in letter writing.
- **D.** It has changed how people connect and interact with each other.

Question: 3 of 30

According to the United Nations Conference on Trade and Development (UNCTAD), which country in the Gulf region has the highest proportion of mobile phone users?

- **A.** Saudi Arabia
- **B.** Oman
- **C.** Kuwait
- **D.** UAE

Question: 4 of 30

What did the cohort study in Denmark find regarding the relationship between cell phone use and brain tumors?

- **A.** There is a significant association between the two.
- **B.** There is no association between the two.
- **C.** There is a weak association between the two.
- **D.** There is a potential association between the two.

Question: 5 of 30

According to the FDA, how can individuals reduce their exposure to radiofrequency energy when using a cell phone?

- **A.** Use a hands-free device
- **B.** Use the phone for shorter conversations or when a landline is available
- **C.** Both A and B
- **D.** Neither A nor B

Direction for Questions: 6 to 9

Read the text and then answer the questions.

Marsian life might have been its own worst enemy

Despite the fact that we know early Mars was warmer, wetter, and more habitable than the present-day freeze-dried arid planet, scientists have yet to discover direct evidence that life ever existed on the Martian surface. Important inquiries persist if Mars was once inhabited by life. What were the consequences of life on the planet, and where might we locate indications of its former existence? Contrarily, a recent investigation into these enigmas concludes that a viable Martian biosphere might have played a significant role in the planet's current inhospitable condition. Certain regions of Mars, including Jezero Crater, where NASA's Perseverance rover is currently deployed, appear to be the most favorable locations on the planet for the quest for life. Furthermore, there is a sinister undertone that suggests life on worlds across the universe might be its own worst enemy.

By re-creating Mars four billion years ago using climate and terrain models, French researchers have reached the conclusion that microorganisms may have flourished mere centimeters beneath a significant portion of the Red Planet's surface, shielded from intense cosmic radiation by the soil that covered the planet. However, that subterranean biosphere would have inevitably receded further into the planet, possibly to its demise, due to self-induced frigid temperatures. The study, which was published in Nature Astronomy, posits that these ancient organisms produced methane by consuming hydrogen and carbon dioxide from the Martian atmosphere. Because each of the three substances can function as a greenhouse gas that traps heat, variations in their relative abundances can have substantial impacts on the surface temperature of a planet. The hypothetical "methanogen" biosphere's net reduction in atmospheric greenhouse gases would have precipitated a global cooling event that resulted in the encasement of the majority of Mars' surface in ice, thereby contributing to the planet's current state of inhospitability and desolation.

In 1965, the late chemist James Lovelock, who was a researcher at NASA's Jet Propulsion Laboratory at the time, formulated a practical approach to identify extraterrestrial life. Lovelock and his colleagues postulated that specific chemical compounds present in the atmosphere of a planet function as biosignatures, signifying the worldwide existence of life. The coexistence of oxygen (from photosynthetic organisms) and methane (from methanogens) on Earth, for example, constitutes a potent biosignature: under ambient conditions, each gas eliminates the other, so the persistence of both gases indicates a steady replenishment that is most easily explained by biological sources. The ongoing scientific pursuit of extraterrestrial life was founded upon the principles established by Lovelock's achievements.

Peter Ward, a paleontologist affiliated with the University of Washington, presented a less sanguine perspective in 2009. Ward contended that life is ultimately self-destructive rather than self-regulating at the planetary scale. As an alternative to the Gaia hypothesis, he derived the name of his concept from the legendary Greek mother Medea, who is said to have murdered her own offspring. In support of his "Medea hypothesis," Ward referenced a number of catastrophic extinctions that occurred throughout Earth's history and suggested that these occurrences might suggest an intrinsic suicidal tendency in life. Over two billion years ago, photosynthetic cyanobacteria released enormous quantities of oxygen into the Earth's atmosphere, which had been virtually devoid of the extremely reactive gas prior to that time. This resulted in the inevitable extinction of the methanogens and other "anoxic" organisms that were resistant to oxygen, which were the planet's former overlords. "A mere examination of Earth's past reveals epochs in which life itself was its own worst enemy," says Ward in reference to the apparent correlation between his Medea hypothesis and Sauterey and colleagues' study. "Moreover, I believe this might have been the situation on Mars as well."

With a characteristically Gaian twist, this calamitous occurrence that put an end to anoxic life on Earth also facilitated the growth of other species: the deluge of atmospheric oxygen was essential for the eventual emergence of the multicellular organisms of our present-day biosphere and the diversification of life on Earth. Determining whether life is ultimately Gaian or Medean may thus be a subjective matter necessitating a broader, interplanetary perspective. However, until life on other worlds is discovered and studied, only speculative comparisons through theoretical studies like Sauterey's are possible. This constraint on the study's findings may indicate a deeper, more fundamental truth. The complex terrestrial biosphere's survival on Mars may have been due to the evolutionary adaptability of microbes on the early Earth, which enabled them to flourish despite otherwise calamitous environmental changes. In contrast, the presumably simpler biosphere on Mars appears to have rapidly disappeared. An ascent toward ever-increasing complexity, according to Ward, could potentially avert a biosphere from a dire Medean fate. He states, "I firmly believe that the only way out—the only way any planet can escape once it acquires life—is to evolve intelligence." According to Ward, technological solutions to mitigate the propensity of life to foul its planetary habitat could only emerge at that time.

The analysis failed to account for the potential presence of contemporary methanogens subterranean on Mars. Although inert geophysical activity could also account for the enigmatic methane plumes that scientists have repeatedly detected in the planet's atmosphere, such a scenario could help explain the plumes.

In the case of ancient Mars, the research identifies specific locations on the planet where the hypothetical microorganisms might have flourished in closer proximity to the surface, thereby placing them within the reach of contemporary fossil investigations. Owing to a worldwide chilling event, these hotspots correspond to uncommon regions of Mars that may have remained devoid of ice for significant portions of the planet's subsequent history, notwithstanding a near-global glaciation. One such location is Jezero Crater, which contains an ancient lake and an expansive sedimentary delta that may have preserved fossils. NASA's Perseverance rover is presently engaged in the retrieval of potentially biosignature-bearing materials for subsequent laboratory analysis on Earth. This is a fortuitous coincidence. However, it is uncertain whether fossil evidence of early methanogens would be accessible in that location. They might be enshrined beneath substantial strata of sediment, impassable by Perseverance.

Regardless of the existence of ancient methanogens on Mars, the findings of the recent research serve as a reminder that life can establish the circumstances for its own success or failure on any celestial body. One-celled organisms possess the capability to render a planet that is otherwise habitable hostile. Additionally, Sauterey ominously continues, "Humans could accomplish that even more rapidly given our technological capabilities."

Question: 6 of 30

In which year do the late chemist James Lovelock—then a researcher at NASA's Jet Propulsion Laboratory—devise a feasible strategy for detecting life on other worlds?

- **A.** In 1965
- **B.** In 1956
- **C.** In 1960
- **D.** In 1985

Question: 7 of 30

Which of the following was Peter Ward's view on life on Mars?

- **A.** Certain chemical compounds in a planet's atmosphere act as so-called biosignatures indicating life's global presence
- **B.** Life is more self-destructive than self-regulating and eventually wipes itself out
- **C.** The flood of atmospheric oxygen proved crucial for our planet's biological diversification
- **D.** None of the above

Question: 8 of 30

Through the article, we can conclude that researchers ...

- **A.** have yet to find any direct proof that life of any kind ever graced the Martian surface.
- **B.** have successfully found proof of life on Mars.
- **C.** have found that Mars is habitable by 2099.
- **D.** have successfully found a new discovery that is secret to human.

Question: 9 of 30

Which of the following describes Mars?

- A. Same as earth
- B. Humans can stay on Mars in 2099
- C. Researchers found signs of life on Mars.
- D. Wetter, warmer, and more habitable than the freeze-dried desert world of today

Direction for Questions: 10 to 14

Read the poem then answer the questions.

Barter
***by* Sara Teasdale**

Life has loveliness to sell,
All beautiful and splendid things,
Blue waves whitened on a cliff,
Soaring fire that sways and sings,
And children's faces looking up
Holding wonder in a cup.

Life has loveliness to sell,
Music like a curve of gold,
Scent of pine trees in the rain,
Eyes that love you, arms that hold,
And for your spirit's still delight,
Holy thoughts that star the night.

Spend all you have for loveliness,
Buy it and never count the cost;
For one white singing hour of peace
Count many a year of strife well lost,
And for a breath of ecstasy
Give all you have been, or could be.

Question: 10 of 30

What is the primary theme of the poem "Barter" by Sara Teasdale?

- A. The importance of material wealth
- B. The value of life's loveliness
- C. The pursuit of happiness
- D. The fleeting nature of beauty

Question: 11 of 30

What poetic device is used in the line "Music like a curve of gold"?

- A. Simile
- B. Metaphor
- C. Alliteration
- D. Onomatopoeia

Question: 12 of 30

In the line "Spend all you have for loveliness," what does the speaker advise the reader to do?

- **A.** Save money for the future
- **B.** Prioritize material possessions
- **C.** Cherish and invest in life's beauty
- **D.** Live a frugal life

Question: 13 of 30

What is the effect of the phrase "holy thoughts that star the night" in the poem?

- **A.** It suggests a religious theme.
- **B.** It conveys a sense of spiritual or transcendent beauty.
- **C.** It emphasizes the darkness of the night.
- **D.** It symbolizes the end of the day.

Question: 14 of 30

In the context of the poem, what does the phrase "never count the cost" mean?

- **A.** Keep track of expenses
- **B.** Ignore the value of beauty
- **C.** Prioritize financial gain
- **D.** Do not hesitate to invest in loveliness

Direction for Questions: 15 to 18

Read the extracts and then answer the questions.

Extract A:
Summer is a season that brings joy and excitement to people all around the world. It is a time when the sun shines brightly, the days are longer, and the weather is warm. It is the perfect season for outdoor activities, such as swimming, hiking, and picnics. Many people go on vacations during this time, exploring new places and creating lasting memories. Summer is also a great time to relax and unwind, whether it's lying on the beach or reading a book in the shade. The vibrant colors of nature, the sweet smell of flowers, and the taste of refreshing fruits make summer a truly delightful season. It is a time to embrace the beauty of nature and enjoy the simple pleasures of life.

Extract B:
Winter is a season that brings a sense of magic and wonder. It is a time when the world transforms into a winter wonderland, with snow-covered landscapes and cozy nights by the fire. The air is crisp, and the sky is clear, creating a serene atmosphere. Winter is a season for warm sweaters, hot chocolate, and spending time with loved ones. The holiday season fills the air with joy and laughter, as families come together to celebrate. Winter also offers unique activities, such as skiing, ice skating, and building snowmen. It is a time to embrace the cold and find beauty in the simplicity of nature. Winter may be cold, but it warms the heart and brings a sense of unity and happiness.

Extract C:
Spring is a season that represents renewal and growth. It is a time when nature awakens from its winter slumber, and the world is filled with vibrant colors and sweet scents. The weather becomes milder, and the days start to get longer, bringing a sense of anticipation and excitement. Spring is a season of new beginnings, as flowers bloom, birds sing, and animals come out of hibernation. It is a time for fresh starts, whether it's planting a garden or cleaning out the house. Spring also brings a sense of joy and happiness, as people shed their winter coats and spend more time outdoors. It is a season that reminds us of the beauty and resilience of nature, and the endless possibilities that lie ahead.

Extract D:
Autumn, also known as fall, is a season that captivates with its beauty and melancholic charm. It is a time when the leaves start to change color, painting the world in hues of orange, red, and yellow. The air becomes crisp, and the days get shorter, creating a cozy and nostalgic atmosphere. Autumn is a season for harvest, as farmers gather their crops and prepare for the colder months ahead. It is a time for warm sweaters, bonfires, and pumpkin-spiced everything. The sound of leaves crunching underfoot, the smell of woodsmoke in the air, and the taste of freshly baked apple pie evoke feelings

of comfort and contentment. Autumn is a season that reminds us of the beauty in change and the cyclical nature of life. It is a time to embrace the fleeting moments and find joy in the simple pleasures of the season.

Question: 15 of 30

Which of the extracts refers to the time when leaves start to change color ?

- **A.** Extract A
- **B.** Extract B
- **C.** Extract C
- **D.** Extract D

Question: 16 of 30

Which of the extracts refers to the cozy nights by the fire?

- **A.** Extract A
- **B.** Extract B
- **C.** Extract C
- **D.** Extract D

Question: 17 of 30

Which of the extracts states about the season when the people go on vacations?

- **A.** Extract A
- **B.** Extract B
- **C.** Extract C
- **D.** Extract D

Question: 18 of 30

Which of the extracts refers to the season which reminds us of the cyclical nature of life?

- **A.** Extract A
- **B.** Extract B
- **C.** Extract C
- **D.** Extract D

Direction for Questions: 19 to 25

Seven sentences have been removed from the text. Choose from the sentences (A – H) the one which fits each gap (1 – 7). There is one extra sentence which you do not need to use.

Tornadoes: Nature's Fierce and Unpredictable Forces

1.... These violent, swirling columns of air can leave behind a path of destruction and devastation, yet they also captivate us with their raw power and beauty. In this essay, we will explore the causes, impacts, and safety measures associated with tornadoes.

Tornadoes typically form from severe thunderstorms, where warm, moist air collides with cold, dry air. **2....** As the rotating column descends from the base of a storm cloud, it can stretch towards the ground, forming a tornado.

3.... The most common type is the "classic" tornado, which appears as a narrow, funnel-shaped cloud extending from the base of a thunderstorm. Other types include multiple vortex tornadoes, which have several small, rotating funnels, and waterspouts, which form over bodies of water.

Tornadoes are known for their destructive potential. **4....** They can go by tearing apart buildings, uprooting trees, and flinging debris with incredible force. The damage caused by tornadoes can be devastating, leaving communities in ruins and causing significant loss of life.

As tornadoes are unpredictable, it is crucial to be prepared and take necessary precautions. **5....** This includes identifying a safe location in your home, such as a basement or an interior room on the lowest level. Additionally, keeping an emergency kit stocked with essentials like water, food, and a battery-powered radio is essential. Staying informed through local weather alerts is also crucial.

Tornadoes are common in an area known as Tornado Alley, which stretches from the central United States to parts of Canada and Mexico. **6...** Advancements in technology have greatly improved our understanding of tornadoes and our ability to predict them. **7....** Additionally, storm-chasing teams and research institutions study tornadoes up close, collecting valuable data to enhance our knowledge of these powerful forces of nature.

A. ***This collision creates an unstable atmosphere, leading to the development of a rotating column of air***
B. ***Tornadoes come in various shapes and sizes***
C. ***The best way to stay safe is to have a plan in place***
D. ***This region experiences a high frequency of tornadoes due to the interaction of warm, moist air from the Gulf of Mexico and dry, cold air from the Rocky Mountains.***
E. ***They can reach wind speeds of over 300 miles per hour***
F. ***Tornadoes are deadly phenomenon***
G. ***Tornadoes are one of the most fascinating and awe-inspiring phenomena found in nature***
H. ***Doppler radar systems can detect the rotation within storms, allowing meteorologists to issue timely warnings.***

Question: 19 of 30

Which one fits in (1) ?

- ○ **A.** Sentence A
- ○ **B.** Sentence B
- ○ **C.** Sentence G
- ○ **D.** Sentence H

Question: 20 of 30

Which one fits in (2) ?

- ○ **A.** Sentence A
- ○ **B.** Sentence B
- ○ **C.** Sentence C
- ○ **D.** Sentence D

Question: 21 of 30

Which one fits in (3) ?

- ○ **A.** Sentence A
- ○ **B.** Sentence D
- ○ **C.** Sentence B
- ○ **D.** Sentence E

Question: 22 of 30

Which one fits in (4) ?

- ○ **A.** Sentence A
- ○ **B.** Sentence B
- ○ **C.** Sentence H
- ○ **D.** Sentence E

Question: 23 of 30

Which one fits in (5) ?

- **A.** Sentence A
- **B.** Sentence B
- **C.** Sentence C
- **D.** Sentence D

Question: 24 of 30

Which one fits in (6) ?

- **A.** Sentence A
- **B.** Sentence B
- **C.** Sentence C
- **D.** Sentence D

Question: 25 of 30

Which one fits in (7) ?

- **A.** Sentence A
- **B.** Sentence G
- **C.** Sentence H
- **D.** Sentence C

Direction for Questions: 26 to 30

Read the poem then answer the questions.

The House on the Hill
***by* EA Robison**

They are all gone away,
The House is shut and still,
There is nothing more to say.

Through broken walls and gray
The winds blow bleak and shrill:
They are all gone away.

Nor is there one to-day
To speak them good or ill:
There is nothing more to say.

Why is it then we stray
Around the sunken sill?
They are all gone away,

And our poor fancy-play
For them is wasted skill:
There is nothing more to say.

There is ruin and decay
In the House on the Hill:
They are all gone away,
There is nothing more to say.

Question: 26 of 30

What is the predominant tone of the poem "The House on the Hill"?

- A. Joyful
- B. Melancholic
- C. Excited
- D. Optimistic

Question: 27 of 30

What is the significance of the repeated line, "There is nothing more to say," in the poem?

- A. It emphasizes the abundance of words spoken.
- B. It highlights the finality and emptiness of the situation.
- C. It suggests that there are unresolved issues.
- D. It indicates the presence of ongoing conversations.

Question: 28 of 30

What deeper meaning the House on the Hill symbolize in the poem?

- A. A place of happiness and celebration
- B. A symbol of eternal life
- C. The passage of time and the inevitability of change
- D. A representation of endless possibilities

Question: 29 of 30

How does the poet convey a sense of abandonment in the poem?

- A. Through vivid descriptions of the house's interior
- B. By emphasizing the presence of people inside the house
- C. Through the use of repetition and the line "They are all gone away"
- D. By portraying the house as a welcoming and lively place

Question: 30 of 30

What does the phrase "our poor fancy-play" in the poem imply?

- A. The imagination's ability to create stories
- B. The importance of creativity in life
- C. The futility of artistic endeavors
- D. The joy of childhood games

Preptive Prepare Better

Series A Mathematical Reasoning Trial Test

Name: ______________________________

Date: ______________________________

Question: 1 of 35

The difference between the present ages of Arun and Deepak is 14 years. Seven years ago the ratio of their ages was 5:7 .
What is Deepak's present age?

- **A.** 49 years
- **B.** 42 years
- **C.** 63 years
- **D.** None of these

Question: 2 of 35

Robert walks slower than Cooper and Cooper walks as fast as Collin and Alan walk faster than Collin.
Who walks the fastest?

- **A.** Cooper
- **B.** Alan
- **C.** Robert
- **D.** Both Cooper and Collin

Question: 3 of 35

There are 4 red buttons, 6 white buttons and 3 black buttons in an invisible cord.
What is the minimum number of buttons that must be drawn one at a time to draw **at least one black button everytime** without looking at the button?

- **A.** 6
- **B.** 8
- **C.** 11
- **D.** 15

Question: 4 of 35

□ represents one number.

△ represents another different number.

If :
7 X □ = 112
△ + □ = 96

What is △?

- **A.** 10
- **B.** 20
- **C.** 42
- **D.** 80

Question: 5 of 35

How many minimum straight lines should be drown inside the figure to make **six small same-size rectangles** ?

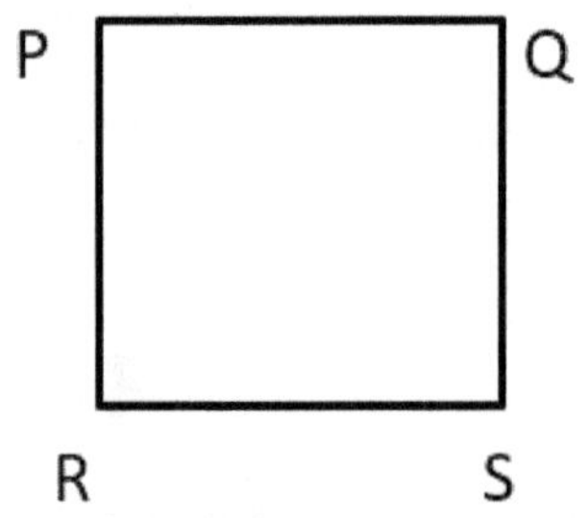

A. 6

B. 3

C. 2

D. 4

Question: 6 of 35

In a hall, there are 85 rows of chairs and each row has 24 chairs. Two of those rows of chairs are brought into an empty classroom and to be re-arranged into rows of 8. How many rows are there in the classroom now?

A. 48

B. 8

C. 9

D. 6

Question: 7 of 35

Khairul and Jason had a total of 1232 cards. After Khairul gave 342 cards to Jason, Khairul still had 298 more cards than Jason.
How many cards did Jason have at first?

A. 250

B. 125

C. 225

D. 230

Question: 8 of 35

There was an equal number of ping pong balls in Box A and Box B at first.
After 47 ping pong balls were removed from Box B and 47 ping pong balls were added in Box A, there were 236 ping pong balls in the two boxes altogether.
How many ping pong balls were in Box B in the end?

A. 156

B. 165

C. 71

D. 175

Question: 9 of 35

There were 20 400 registered voters for a certain election. 2% of them did not come for voting.
How many voters did come for voting?

- A. 16 992
- B. 20 000
- C. 15 598
- D. 19 992

Question: 10 of 35

Jenny earned \$3 for every box of food that she delivered. She received an additional \$20 for every 10 boxes of food delivered. How many boxes of food did she deliver if she earned a total of \$218?

- A. 22
- B. 120
- C. 9
- D. 46

Question: 11 of 35

A baker bought some eggs. She used $\frac{1}{3}$ of the eggs to bake tarts, 92 eggs to bake muffins and $\frac{2}{3}$ of the remaining eggs to bake cupcakes. After that, she still had 36 eggs. How many eggs did the baker buy?

- A. 525
- B. 309
- C. 300
- D. 209

Question: 12 of 35

During a sale, cooking pots and frying pans were sold at the prices shown in the table below.

Cooking Pot	Frying Pan

Buy 1 at \$810.90	Buy 1 at \$640
Buy 2 or more at \$800 each	Buy 2 or more at \$630.50 each

Naomi bought 1 cooking pot and 4 frying pans during the sale. What was the smallest amount of money she paid?

- A. \$4354.60
- B. \$3332.90
- C. \$3342.80
- D. \$5445.10

Question: 13 of 35

Kelly woke up 20 minutes earlier than usual. As she woke up early, the time she took to get ready is delayed by 6 minutes. Usually, she arrives at school exactly at 7.05 am.
Mcki woke up 10 minutes later than usual. Usually, he arrives at school 18 minutes earlier than Kelly.
Calculate the time difference between Kelly and Mcki arriving at school that day.

- **A.** 6 minutes
- **B.** 5 minutes
- **C.** 10 minutes
- **D.** 7 minutes

Question: 14 of 35

Number 85 can be split in two perfect squares into two different ways. That is: 85 = 81 + 4 or 85 = 49 + 36.
Select another number in the given choices that is less than 100 and can be split into two different squares.

- **A.** 36
- **B.** 40
- **C.** 25
- **D.** 72

Question: 15 of 35

The mean temperature for 7 days is 24.60°C. For the first six days, the mean temperature was 21.60° C. What was the temperature on the 7th day?

- **A.** 42.60°C.
- **B.** 36.60°C.
- **C.** 24.60°C.
- **D.** 21.60°C.

Question: 16 of 35

May had three times as much money as June. May spent $1700. As a result, she had $300 less than June. How much money did May have left?

- **A.** $449
- **B.** $440
- **C.** $350
- **D.** $400

Question: 17 of 35

Miss Sharma baked some pies to be packed into boxes.
If she packed 5 pies in each box, she would have 4 pies left. If she packed 9 pies in each box, she would have 8 pies left.
Given that Miss Sharma baked fewer than 60 pies, how many pies did Miss Sharma bake?

- **A.** 44
- **B.** 48
- **C.** 54
- **D.** 24

Question: 18 of 35

Karen went to the post office to send some letters to her friends in Japan. The postage rates to Japan is shown In the table below.

Mass of letter	
First 30 g	$1.50
Every additional 10 g or part thereof	40 ¢

Karen wanted to send a letter with a mass of 55 g to Japan. She gave the cashier a $5 note. How much change did she get?

- **A.** $4.00
- **B.** $2.30
- **C.** $3.50
- **D.** $6.50

Question: 19 of 35

A small square of side 6 cm is cut out from a bigger square of area 144 cm^2. Find the ratio of the shaded area to that of the unshaded area in the simplest form.

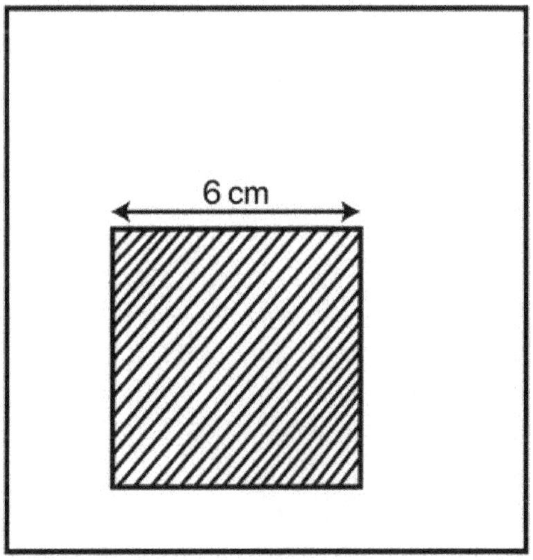

- **A.** 1:5
- **B.** 3:1
- **C.** 1:3
- **D.** 2:5

Question: 20 of 35

There are some marbles in a container. $\frac{1}{3}$ of the marbles are blue. $\frac{1}{4}$ of the remaining marbles are yellow and the rest are red.
What fraction of the marbles in the container are yellow?

- **A.** 6
- **B.** $^1/_6$
- **C.** $^1/_3$
- **D.** $^7/_{16}$

Question: 21 of 35

Terry read $\frac{1}{5}$ of the number of pages in a book in the first hour, $\frac{3}{8}$ of the remaining pages in the second hour. After that, he still had 45 more pages to read.
How many pages are there in the book?

- **A.** 19
- **B.** 90
- **C.** 15
- **D.** 99

Question: 22 of 35

The pie chart shows the results of a survey conducted to identify the favorite game of some students.

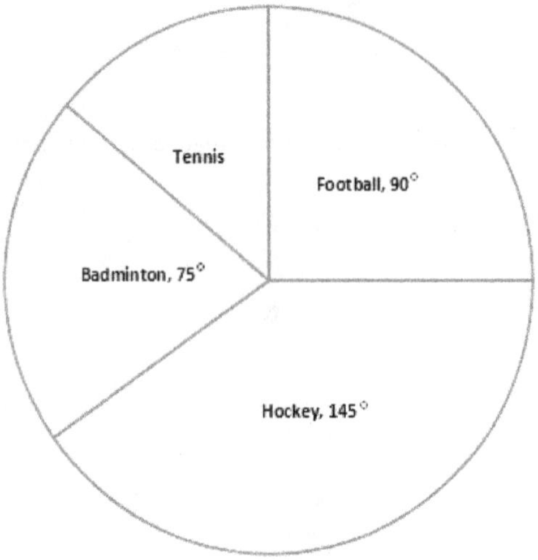

How many students like Tennis if the total number of students is 1,260?

- **A.** 175
- **B.** 90
- **C.** 155
- **D.** 65

Question: 23 of 35

The sides of a triangle are in the ratio : $\frac{1}{2} : \frac{1}{3} : \frac{1}{4}$

If the perimeter of the triangle is 52 cm, what is the length of the shortest side?

- **A.** 9 cm
- **B.** 10 cm
- **C.** 11 cm
- **D.** 12 cm

Question: 24 of 35

Tomek, Romek, Andrew, and Michael said the following about a certain number:

***Tomek**: "This number is equal to 9."*
***Romek**: "This number is prime."*
***Andrew**: "This number is even."*
***Michael**: "This number is equal to 15."*

Only one statement given by either Romek or Tomek is true,
Only one statement given by either Andrew or Michael is true.
What number is it?

- **A.** 2
- **B.** 3
- **C.** 9
- **D.** 15

Question: 25 of 35

The total cost of a pen and 2 similar highlighters is $6.50, The total cost of 2 such pens and 1 such highlighter is $7.
What is the cost of 5 such highlighters?

- **A.** $13
- **B.** $10
- **C.** $9
- **D.** $6

Question: 26 of 35

Aaron has $\frac{3}{5}$ as much savings as Bob. If Bob gives $40 to Aaron, they will have the same amount of money.
How much money does Bob have?

- **A.** $350
- **B.** $225
- **C.** $200
- **D.** $180

Question: 27 of 35

The bar graph below shows the number of old textbooks donated by five Primary 5 classes to a children's home.

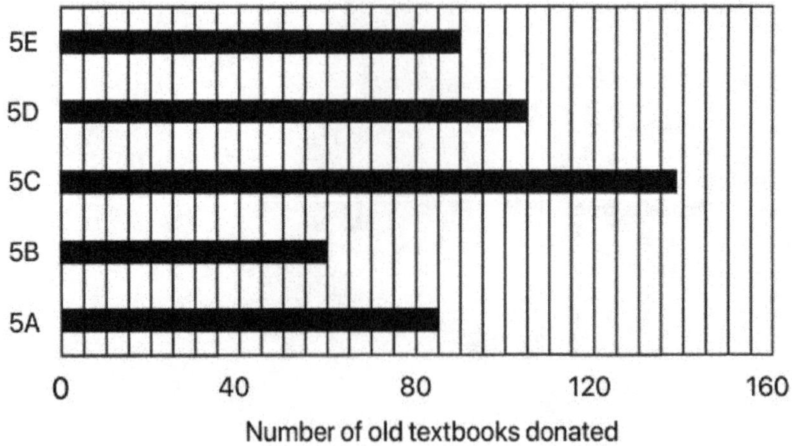

What fraction of the old textbooks was donated by class 5E? Give your answer in the simplest form.

- **A.** $^{5}/_{16}$
- **B.** $^{3}/_{7}$
- **C.** $^{3}/_{16}$
- **D.** $^{3}/_{8}$

Question: 28 of 35

In the figure, ABCD is a rectangle where AC and BD are straight fines. Given that ∠AEB = 130° and AE = EC = DE = EB, find ∠ADB.

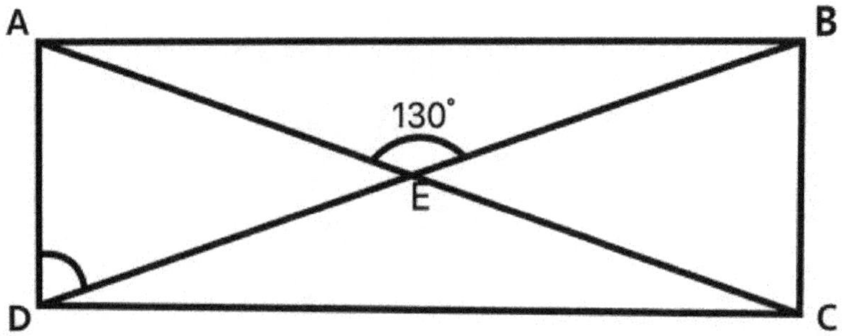

- **A.** 75°
- **B.** 65°
- **C.** 130°
- **D.** 50°

Question: 29 of 35

@, #, $ represent 3 numbers.

#-$=7
@/6+$=#-2
$ ×11=121

What is the value of @?

A. 18

B. 24

C. 30

D. 42

Question: 30 of 35

A large rectangle has side lengths of 11 cm and 21 cm. Three identical rectangles of length 5 cm and width 2 cm are removed. It leaves the below figure.

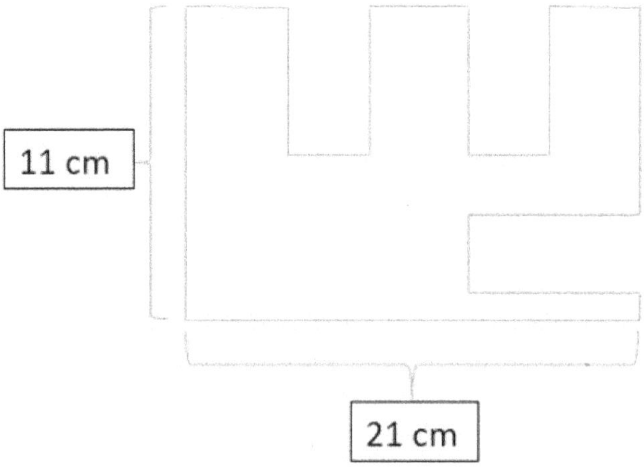

What is the perimeter of the given figure?

A. 64cm

B. 76cm

C. 88cm

D. 94cm

Question: 31 of 35

Some money was given to three daughters A, B and C by their mother to buy household things. The total amount of money given was $4350.
A spent $^{6}/_{15}$ and B spent $^{17}/_{30}$ of the total amount and C spent the remaining.

Which of the following statement(s) is/are correct?

1. The highest amount was spent by A.
2. Amount spent by C was $^{1}/_{30}{}^{th}$ of the total amount.
3. Amount spent by B was more than 2 quarters of the total amount.

A. Statement 1 only

B. Statement 2 only

C. Statements 1 and 2 only

D. Statements 2 and 3 only

Question: 32 of 35

50	21	12	A
62	52	81	10
28	39	95	43
80	11	03	111

Identify the pattern and find A.

- **A.** 64
- **B.** 88
- **C.** 111
- **D.** 122

Question: 33 of 35

Arthur and Merlin did the game of throwing an unbiased cube shaped dice.
Arthur threw the dice first and got number 3. Then Merlin threw the dice once.

Which of these statement(s) is/are correct?

A. *The probability of Arthur getting 3 is* $^{1}/_{6}$.
B. *The probability of Merlin getting a number greater than Arthur is* $^{1}/_{2}$.
C. *The probability that Merlin and Arthur's scores having a difference less than 3 is* $^{5}/_{6}$.

- **A.** Only A
- **B.** Only A and B
- **C.** Only B and C
- **D.** All A, B and C

Question: 34 of 35

The following pie chart shows unsold food (in Kgs) in the canteen at close of business.

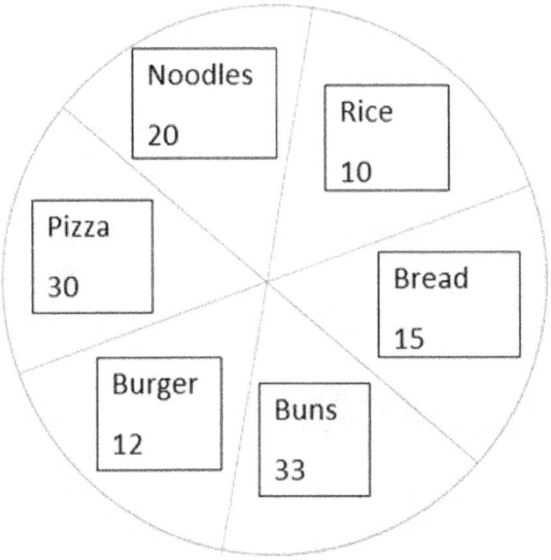

What should be the angle at the center for burger and pizza respectively?

- A. 90°, 36°
- B. 45°, 60°
- C. 36°, 90°
- D. 60°, 90°

Question: 35 of 35

The below bar graph shows favourite books of students in a high school.

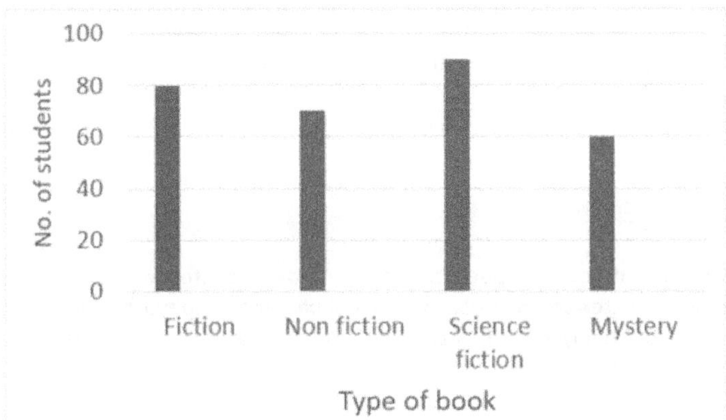

If above information were shown in a pie chart, what would be the difference of the angles at the center of fiction and mystery books?

- A. 45°
- B. 38°
- C. 24°
- D. 16°

Name: ______________________________

Date: ______________________________

Question: 1 of 40

In a railway station, there is a bulb indicator system for train arrivals and departures. Whenever the train leaves the nearest station, the yellow bulb turns on to indicate that the train will arrive soon. And when the train is going to leave the station within 10 minutes, the red light starts to flash on and off.

Richard: "The red bulb is on continuously, which means that the train has already left. We have to wait for another train"

Samson: " The yellow bulb turned on, which means another train will arrive soon. Let's wait."

If the given information is correct, whose reasoning must be correct?

- **A.** Richard only
- **B.** Samson only
- **C.** Both Richard and Samson.
- **D.** Neither Richerd nor Samson.

Question: 2 of 40

Parasitic plants are plants that depend on other plants for nutrients. All parasitic plants are divided into two types. Partial parasites are those that depend on the host plant only for water and minerals while total parasites depend on the host plant for water, minerals and food too. Cuscuta which is commonly known as Dodder is a parasitic plant. Since they lack chlorophyll, they cannot use photosynthesis to produce own food.

If the given information is correct, which one of the following sentences is **not correct**?

- **A.** Cuscuta must be a total parasitic plant.
- **B.** Cuscuta plants depend on the host plant only for water and minerals.
- **C.** Cuscuta plant depends on the host plant for all water, minerals and food.
- **D.** Partial parasites don't absorb food from the host plant.

Question: 3 of 40

Whoever made the choreography of the first-place winning dance must have unique ideas and a keen passion.

If this is true, which one of these statements must also be true?

- **A.** If Liam has unique ideas and a keen passion, he must be the choreographer of the first-place-winning dance.
- **B.** Liam is a choreographer, and if he has unique ideas and a keen passion then he must be the person who made the choreography of the first-place winning dance
- **C.** If Liam is not the person who made the choreography of the first-place-winning dance, he cannot have unique ideas.
- **D.** If Liam doesn't have unique ideas, he cannot be the choreographer of the first winning dance.

Question: 4 of 40

In a team game, they have to draw lots first. There are two teams. The team who got blue colour lot have to swim through the lake first and then have to follow the route and find the hidden map. The team members who got the red lot first had to go through the mud pit and then find the hidden map. Both hidden maps show the path to the same destination. The team that reaches the destination first will win the game.

Team A got the red lot, but they lost the game.

Which of the following **cannot** be the reason?

- **A.** It was more difficult to go through the mud pit than swim across the lake.
- **B.** They took a lot of time to follow the hidden map and find the destination.
- **C.** They took a lot of time to swim across the lake.
- **D.** Team B did their activities very quicker than team A.

Question: 5 of 40

If Sally wakes up late she will skip her breakfast and go to school.
If Sally skips her breakfast she will faint during the morning assembly.
Whenever a student faints during the assembly in the school they have to rest in the sick room.

Samuel: "If Sally faint during the morning assembly then she must have skipped her breakfast."

Raven: "If Sally rests in the sick room, then she must have woke up late."

If the information above is true, whose reasoning must be correct?

- **A.** Samuel only
- **B.** Raven only
- **C.** Both Samuel and Raven
- **D.** Neither Samuel nor Raven

Question: 6 of 40

People with choleric personalities are likely to engage in risky behaviour. However, all of them are short-tempered when they are under stress.

Henry: "I know Ben has a choleric personality so he will not think twice to engage in risky behaviours"

Alexa: "Nyla is short-tempered when she is under stress. She must have a choleric personality."

If the information in the box is true, whose reasoning must be correct?

- **A.** Henry only
- **B.** Alexa only
- **C.** Both Henry and Alexa
- **D.** Neither Henry nor Alexa.

Question: 7 of 40

The book "Rich dad poor dad" which is written by Robert Kiyosaki and Sharon Lechter is a great book, and an inspiration for anyone who wants to improve his or her understanding of how money works.

Kim: "I read a book last year. Its cover page was damaged. At that time I was bankrupt and that book was a great inspiration for me to pull myself together and become successful"

Joseph: "Oh, then that book must be Rich dad poor dad."

Which one of these sentences shows the mistake Joseph has made?

- **A.** Reading a book cannot help anyone to overcome the obstacles in their lives, the only solution is hard work.
- **B.** "Rich dad poor dad" is a very old book which doesn't fit the modern financial background, so nowadays no one reads that book.
- **C.** Joseph doesn't know about Kim's financial history.
- **D.** There might be many other books which inspire people who want to improve their understanding of how money works.

Question: 8 of 40

The founder of the SGH construction company owns a privet jet and four luxurious cars.

If the information is true, which one of the following must be true?

- **A.** William owns a private jet and four luxurious cars, therefore he must be the founder of the SGH construction company
- **B.** Simon owns a privet jet and also he is a founder of a company too therefore he must be the founder of SGH construction company.
- **C.** Benson owns a privet jet and only two luxurious cars, therefore he cannot be the founder of the SGH construction company.
- **D.** Fredric has four luxurious cars. He cannot be the founder of the SGH construction company.

Question: 9 of 40

If Julia changes her old telephone number, her old friends will not be able to contact her via telephone.
If Julia's old friends are unable to contact her, she will not get the invitation to their annual get-together.
Julia will go to the get-together only if she gets the invitation.
If Julia is going to the get-together she will likely buy a new gown.

If the above information is correct, which one of the following is not possible?

- **A.** Julia changed her old telephone number but she went to the get-together.
- **B.** Julia got the invitation but she didn't buy a new gown.
- **C.** Julia's friends weren't able to contact her but she went to the get-together.
- **D.** Julia didn't change her telephone number but she didn't go to the get-together.

Question: 10 of 40

The blobfish is a pink, marine water fish that grows up to a foot long, and found in the deep waters off the coasts of Australia and Tasmania. Blobfish do not have any teeth. The blobfish is often caught in trawling nets by fishermen.

Based on above, which one is most likely to be wrong ?

- **A.** A fisherman found a pink fish off the coasts of Australia but it was only half a foot long therefore it can be a blobfish.
- **B.** A fisherman found a foot-long pink fish in a river, it might be a blobfish.
- **C.** A fisherman found a foot-long pink fish with very tiny teeth off the coasts of Tasmania, it cannot be a blobfish.
- **D.** A fisherman found a blobfish, so it must be a foot long.

Question: 11 of 40

A nurse in a blood donating campaign told: "To donate blood the donator's weight must be more than 50kg and he/she be over 18 years old. You cannot donate if you have a cold, flu, sore throat, or any other infection. Also if you have recently had a tattoo or body piercing you cannot donate for 6 months from the date of the procedure."

Noah is in his late twenties. Noah does not have any piercings. Nine mounts ago he got his last tattoo.

If the above information is correct which of the following situation is a **certainty**?

- **A.** Even though Noah is healthy and weighs more than 50kg, he cannot donate his blood because of his tattoo.
- **B.** Noah wasn't able to donate his blood because he is too young.
- **C.** Noah's age and weight are 27 years and 65kg. He doesn't have any health problems. He would surely not be able to donate his blood.
- **D.** Noah has a sore throat and flu today so he definitely can't donate his blood.

Question: 12 of 40

Sports teacher observed: *All the students in Adam's class are better cricket players than some students in David's class.*

Which of the following is **opposite** to sport teacher's observation ?

- **1.** There is at least one student in Adam's class who is a terrible cricket player than all of the students in David's class.
- **2.** All the students in Adam's class are terrible cricket players.
- **3.** All the students in David's class are better cricket players than all the students in Adam's class.
- **4.** Students in both classes are terrible players.

Question: 13 of 40

I have opened today's newspaper and glanced through the headlines. These two caught my attention.

A: There is a low atmospheric pressure region formed.
B: The airport announced some flight cancellations.

What is the logical relation between A and B?

- **1.** A is the effect and B is the cause.
- **2.** A is the cause and B is the effect.
- **3.** A and B are independent.
- **4.** None of the above.

Question: 14 of 40

Emma and 9 other students are sitting in a circle participating in a game.
Starting from Emma, they are counting 1, 2, 3, 1, 2, 3, ... around the circle.
Which number will Emma say in the second round?

A. 1
B. 2
C. 3
D. 4

Question: 15 of 40

If the 1st of May is a Sunday, when is the last Sunday of the month?

A. 29th
B. 22nd
C. 30th
D. 31st

Question: 16 of 40

Whenever Justin uses his motorcycle, Justin arrives early at school.
When Justin arrives early at school, Justin has his breakfast at the school canteen.

Logan: "Justin arrived early at school today; he must have used his motorcycle on his way here."

Paul: "Justin is having his breakfast at the school canteen; he must have used his motorcycle on his way here."

If the information is true, whose reasoning is correct?

A. Logan only
B. Paul only
C. Both Logan and Paul
D. Neither Logan nor Paul

Question: 17 of 40

Whenever Joe has a runny nose, he takes anti-histamine. When he takes anti-histamine, he falls asleep.

Lester: "Joe took an anti-histamine, he must have a runny nose."

Jane: "If Joe have a runny nose then he'll fall asleep."

If the information is true, whose reasoning is correct?

- **A.** Lester only
- **B.** Jane only
- **C.** Both Lester and Jane
- **D.** Neither Lester nor Jane

Question: 18 of 40

Tourmaline aids in the formation of a barrier around a person or space to keep bad or unwanted energies out. It is also grounding and beneficial for chakra balance. The stone can even eliminate negative thinking patterns and challenging energy, transforming them into more positive energy and beliefs. Tourmaline, as a therapeutic crystal, helps you understand yourself and others, boosts self-confidence, and reduces fear. It draws motivation, compassion, tolerance and wealth. Tourmaline, as a potent mental healer, balances the right and left hemispheres of the brain and transforms negative energy into good.

If the information is true, which of the following must be false?

- **A.** Tourmaline is a healing crystal that fosters self-confidence, reduces fear, and aids in understanding oneself and others.
- **B.** Tourmaline stone attracts negative energy to come towards you resulting from any form of disturbance.
- **C.** Tourmaline balances the right and left hemispheres of the brain and transforms negative energy into positive as a potent mental healer.
- **D.** Tourmaline grounds you and aids in bringing each chakra into balance.

Question: 19 of 40

When you have anemia, your body doesn't produce enough healthy red blood cells to supply your tissues with enough oxygen. Being anemic, or having low hemoglobin, can make you feel exhausted and frail. Fatigue, weakness, pale or yellow skin, irregular heartbeats, shortness of breath, lightheadedness, chest pain, cold hands and feet, and headaches are all indications of anemia.

If the information is true, which of the following must be false?

- **A.** Pale skin alone is the symptoms of having an anemia.
- **B.** Anemia is a disorder in which there are insufficient healthy red blood cells to transport oxygen to your body's tissues.
- **C.** Anemia, commonly known as low hemoglobin, can cause fatigue and weakness.
- **D.** Fatigue, weakness, pale or yellowish complexion, irregular heartbeats, shortness of breath, dizziness or lightheadedness are among the signs and symptoms of having an anemia.

Question: 20 of 40

Those who suffer from spectrophobia may have intense phobias of their own reflection, the mirror itself, or spirits that may show up in mirrors. Despite being extremely uncommon, this condition can also be quite serious. Spectrophobia, like other phobias, can cause disruptions in many facets of a person's life and cause avoidance behaviors.

Benson: "Loisa's hair and make-up was done by her in laws. She doesn't want to even see herself in the mirror after her hair and make-up was done. She must have a spectrophobia."

Cathy: "Johan's sister doesn't like to look in the mirror. His sister fears that something might suddenly show up in the mirror. His sister may have a spectrophobia, who knows."

Whose reasoning is correct?

- **A.** Benson
- **B.** Cathy
- **C.** Benson and Cathy
- **D.** Neither Benson nor Cathy

Question: 21 of 40

The plant compounds flavonoids, which have anti-inflammatory and antioxidant properties, are abundant in onions. These substances may offer defense against long-term diseases including cancer and diabetes when eaten often and in appropriate amounts. Eating onions can help keep your cardiovascular system in good shape. Onions contain a polyphenol called quercetin, which lowers lipid and high blood pressure levels.

If the information is true, which of the following is false?

- **A.** Flavonoids found in onions may be a defense against cancer and diabetes.
- **B.** Onions contains polyphenol that are not good for the heart.
- **C.** Onions must still be eaten in appropriate amounts.
- **D.** The quercetin found in onions are found to lower lipid and high blood pressure.

Question: 22 of 40

Statement 1: "Only students who score above 85% on the physics test can attend the field trip."
Statement 2: "Alex scored 88% on the physics test."

Natalie: "Alex will surely attend the field trip because he scored above 85% on the physics test, and only those who do can go."
Liam: "Alex can only attend the field trip if he scored above 85% on the physics test."

If the information in the bold letters is true, whose reasoning is correct?

- **A.** Natalie Only
- **B.** Liam Only
- **C.** Both Natalie and Liam
- **D.** Neither Natalie nor Liam

Question: 23 of 40

A language teacher advises, "Practicing speaking with native speakers is crucial for improving language fluency."

Which one of the following statements, if true, best supports the teacher's advice?

- **A.** Some individuals believe that they can become fluent without speaking to native speakers.
- **B.** Finding native speakers to practice with can be challenging, especially for those in non-English-speaking countries.
- **C.** Advanced language learning apps often replace the need for practicing with native speakers.
- **D.** Speaking with native speakers exposes language learners to authentic pronunciation and conversational nuances, contributing to fluency improvement.

Question: 24 of 40

Only students who complete the advanced writing course are eligible to participate in the writing competition.

Oliver: "Samantha didn't participate in the competition, so she must not have taken the advanced writing course."

Isabella: "Not necessarily, Samantha might have completed the course but chosen not to participate."

If the statement in the bold letters is true, whose reasoning is correct?

- **A.** Oliver only
- **B.** Isabella only
- **C.** Both Oliver and Isabella
- **D.** Neither Oliver nor Isabella

Question: 25 of 40

As Country X celebrates the day it gained its independence, shops, malls, and supermarkets remain closed.

Emmy: Today isn't Independence Day because many shops are open.
Anna: Last Tuesday, I couldn't find a stall to buy groceries since everyone was closed on Independence Day.

If that's the case, who had the correct reasoning?

- **A.** Emmy
- **B.** Anna
- **C.** Both Emmy and Anna
- **D.** Neither Emmy nor Anna

Question: 26 of 40

According to the results of a certain class taking Maths, English, and Crafts, the following conclusions were made :

i. *No one passed all the three tests.*
ii. *No one failed all the three tests.*
iii. *Everyone who passed Maths also passed English.*
iv. *Everyone who passed Craft also passed either Maths or English.*

So, what can't be true about Eric?

- **A.** Eric only passed Maths.
- **B.** Eric passed both Maths and English.
- **C.** Eric passed both Maths and Crafts.
- **D.** Eric passed English and failed in Maths.

Question: 27 of 40

In a class with 4 streams (reading groups), students were asked to choose their preferred one among Neptune, Jupiter, Venus, and Pluto. Out of the 80 students, only 20 chose Neptune. The number of students who chose Jupiter was twice those who settled for Venus. Only 30 students chose Pluto.

Which of the following reasoning is correct?

- **A.** Pluto is the least-loved stream.
- **B.** Neptune is loved more than Jupiter.
- **C.** Venus is the least loved stream.
- **D.** All the streams are loved equally.

Question: 28 of 40

For a team to participate in a certain tournament, its football club must meet at least one of the following conditions.

i. *Become the top 3 among the teams in its qualifying group.*
ii. *Having won at least one trophy in the last season.*
iii. *Gathering at least 20 points in the previous season.*

Which of the following teams **won't participate** in this tournament?

A. Team A, which scored 25 points during the last season and won 2 trophies

B. Team B, which was position 2 in its qualifying group but didn't get a trophy

C. Team C, which hasn't won a trophy for the last 3 seasons and was position 4 in its qualifying group

D. Team D which was the winner of the Team of the Year trophy last season and gathered 30 points during that season

Question: 29 of 40

Since the debate contest started 3 years ago, no one has ever won the trophy twice. Last year, James won the competition, whereas Richie came second. According to this year's result, James didn't break the tradition despite participating in the contest for another time. Richie also gave the contest a second shot but did even worse than he had before. The only other contestants were Janet and Elsa. If Janet took the last position, who won this time around?

A. James

B. Richie

C. Janet

D. Elsa

Question: 30 of 40

There is a school trip that 40 students should attend. The teacher in charge gives chances on a **first-come, first-served** basis. The deadline for applying for this trip is Monday.

Joe: Since I submitted my application before Monday, I will surely be attending the school trip this time.
Joy: I submitted my application on Monday; hence, I won't be attending this school trip.

Given the two statements, who has the correct reasoning?

A. Both Joe and Joy

B. Neither Joe nor Joy

C. Joe only

D. Joy only

Question: 31 of 40

John and Mary like communicating using a code. Therefore, they have devised a way of decoding messages to maintain privacy. The code is as follows:

A	B	C	D	E	F	G	H	I	J	K	L	M	N	O	P	Q	R	S	T	U	V	W	X	Y	Z
1	2	3	4	5	6	7	8	9	0	1	2	3	4	5	6	7	8	9	0	1	2	3	4	5	6

Under such circumstances, which words can share the same code as BOOK?

A. LOOK

B. VASE

C. LYOD

D. BELL

Question: 32 of 40

From a certain location, Agnes walks 70 m towards the south. She then turns right and travels a distance of 70 m straight. She turns left, walking up to 60 m straight. Lastly, she turns left and walks for 70 m straight. So, what's the distance between where she is and where her journey began?

- **A.** 270 m
- **B.** 200 m
- **C.** 130 m
- **D.** 70 m

Question: 33 of 40

If you walk regularly, going to the gym to work out is unnecessary. Equally important, you don't need fitness equipment to exercise. The only thing necessary is comfortable shoes, especially athletic ones.

Which of the following is the best **interpretation** of the above statements?

- **A.** Walking regularly is better than working out.
- **B.** If you don't use athletic shoes, you will hurt your feet when walking.
- **C.** Walking regularly is as effective as working out to keep fit.
- **D.** You should combine walking and working out for fitness.

Question: 34 of 40

It is the year 2000, and Alex has been asked by his class teacher to calculate the number of days in the months with an even number of days. He follows the following steps.

i. **Out of the months with an even number of days, there are February, April, June, September, and November.**
ii. **The number of days are 28, 30, 30, 30 and 30.**
iii. **The sum is 28 + 30 + 30 + 30 + 30=148**

Which of the following is conclusion regarding Alex's approach ?

- **A.** Alex got it right.
- **B.** Alex was bound to fail solving the question after step 1 because step 1 is incorrect.
- **C.** Alex was bound to fail solving the question after step 2 because spep 2 is incorrect.
- **D.** Alex was bound to fail solving the question after step 3 because step 3 is incorrect.

Question: 35 of 40

A train from Town A to B picks 4 people along the road.

i. **Jane isn't the first or the last person to board to train.**
ii. **Only Emily enters after Jane.**
iii. **Agatha enters the train first.**

If the first two statements are true, the third one is ....

- **A.** True
- **B.** False
- **C.** Uncertain
- **D.** None of the above

Question: 36 of 40

If the exchange rate of USD vs Euro is 1.2 USD for 1 Euro, which of the following reasoning is correct?

Mark: If I exchange my USD for Euros, the number of Euros will be relatively high.
Jimmy: Since I will be exchanging my Euros for USD, the USD will be relatively lower.

- **A.** Mark
- **B.** Jimmy
- **C.** Both Mark and Jimmy
- **D.** Neither Mark nor Jimmy

Question: 37 of 40

Different people had various opinions regarding Daniel's weight. **A** said that the weight was between 60 and 70 kg. **B** had a different range of between 50 and 65 kg. **C** said that the weight is less than or equal to 62.
If all the opinions were factual, which of the following conclusions is correct?

- **A.** Daniel's weight can be 65
- **B.** Daniel's weight can also be 68
- **C.** Daniel's weight can be 58
- **D.** Daniel's weight can be 60

Question: 38 of 40

In an exam no two students got the same marks. If you want to find out one student's rank (from top) in the exam, which of the following statements would help you figure it out?

i. ***The class has 50 students.***
ii. ***Only 15 students scored more than the individual in question.***

- **A.** i only
- **B.** ii only
- **C.** both statements together
- **D.** none of the above

Question: 39 of 40

There are three sentences :

Jake read more books than Antonio.
Antonio read fewer books than Abby.
Abby read more books than Jake.

If the first two statements are true, the third one is....

- **A.** True
- **B.** False
- **C.** Insufficient data to connect
- **D.** None of the above

Question: 40 of 40

To avoid idleness in the office, four workers have decided to work in shifts, each working alone in a day, from Monday to Thursday.
The schedule showed that Job (office worker) was to work on Monday. However, he devised an arrangement to switch dates with Leon, who was to work on Wednesday. Ann also traded her working day with Emily, who (Emily) was to work on Thursday.

So, who was on duty on Tuesday, eventually?

- **A.** Job
- **B.** Emily
- **C.** Ann
- **D.** Leon

Preptive Prepare Better

Series B Reading Comprehension Trial Test

Name: ______________________________

Date: ______________________________

Direction for Questions: 1 to 4

Read the text below and answer the questions.

Kansas tornadoes

Kansas is famous for many things, including wheat, sunflowers, and tornadoes!

L. Frank Baum's The Wonderful Wizard of Oz relates the narrative of Dorothy, who is caught in a Kansas storm and ends up in the fictitious realm of Oz. This story has been adapted for the stage and the big screen. You may have seen the film adaptation, starring Judy Garland as Dorothy.

Tornadoes are thunderstorms characterized by rapidly spinning winds that generate a funnel cloud. They are also known as "twisters," since they reach downward from the massive clouds. Winds rotating within a tornado can reach speeds of nearly 300 miles per hour!

A tornado normally rushes across an area rapidly; it can do significant damage. There have been several exceptional tornado reports. In one case, a schoolhouse was demolished while the 85 children who were originally inside were dragged 400 feet with no one harmed. There were also five railway coaches, each weighing 70 tons, that were lifted from their tracks.

Question: 1 of 30

Who is L. Frank Baum?

- **A.** The character in The Wonderful Wizard of Oz children's novel
- **B.** The director of The Wonderful Wizard of Oz movies
- **C.** The author of The Wonderful Wizard of Oz children's novel
- **D.** One of a tornado victim

Question: 2 of 30

What is the other name of a tornado?

- **A.** Twisters
- **B.** Toaster
- **C.** Rotating winds
- **D.** Thunder

Question: 3 of 30

Based on the article above, Kansas is known for many things except ...

- **A.** Tornadoes
- **B.** Tomatoes
- **C.** Wheat
- **D.** Sunflowers

Question: 4 of 30

The winds that rotate within a tornado usually reach a speed of ...

- A. Almost 300 miles per hour
- B. 300-350 miles per hour
- C. More than 300 miles per hour
- D. None of the above

Direction for Questions: 5 to 7

Read the text below and answer the questions.

The Coliseum

The Coliseum is an ancient stadium located in the heart of Rome. It is the most massive of its kind. It is quite old. They began construction in the year 70. It took ten years to construct. It is still in existence today.

The Coliseum has been used in a variety of ways. Men fought each other in it in ancient Rome. They battled lions, tigers, and bears. It was terrifying. However, the majority of individuals enjoyed it. Up to 80,000 Romans would crowd inside to see.

Over the years, the Coliseum has been destroyed numerous times. In the year 217, it was struck by lightning. This set off a fire. The Coliseum is mostly composed of stone. However, the fire caused damage to the top levels. They were constructed of wood. It took many years to remediate this harm. It wasn't completed until the year 240.

The most severe devastation occurred in 1349. A powerful earthquake jolted Rome and the Coliseum. The building's south side fell. The arena's remains were scattered across the ground. Many folks collected the broken stones. Others took stones off the benches. They were used to rebuild buildings and churches.

The Coliseum is now one of Rome's most prominent tourist attractions. People come from all around the world to view it. Every Good Friday, the Pope leads a large march around it. It is a well-known symbol. It's even been seen on the reverse of a coin.

Question: 5 of 30

When did the Romans finish building the Coliseum?

- A. The year 70
- B. The year 523
- C. The year 80
- D. The year 240

Question: 6 of 30

What caused the fire that damaged the upper levels of the Coliseum?

- A. A bolt of lightning
- B. Rowdy people who came to watch the events
- C. An attacking army
- D. An angry mob

Question: 7 of 30

Which caused the most damage to the Coliseum?

- A. Fires
- B. Earthquakes
- C. Wars
- D. Hurricanes

Direction for Questions: 8 to 12

Read the extracts below and answer the questions.

Extract A : What exactly is a migraine?

A migraine headache is characterized by strong throbbing pain or a pulsating sensation on one side of the brain. It is frequently accompanied by nausea, and excessive light and sound sensitivity. Migraine attacks can persist for hours or days, and the pain can be severe enough to interfere with daily tasks.

A warning sensation known as an aura develops before or with the headache in some people. Auras can involve visual abnormalities like flashes of light or blind spots, as well as physical symptoms including tingling on one side of the face, arm, or leg, and difficulties speaking.

Some migraine medications can help prevent migraines and make them less uncomfortable. The correct medications, in conjunction with self-help treatments and lifestyle adjustments, may be beneficial. Migraines, which can afflict both children and adults, can develop through four stages: prodrome, aura, attack, and post-drome. Not everyone who suffers from migraines progresses through all stages.

You may notice subtle changes one or two days before a migraine, such as constipation, mood changes ranging from depression to euphoria, food cravings, neck stiffness, increased urination, fluid retention, and frequent yawning.

Auras can appear before or during migraines for certain people. Auras are reversible nervous system symptoms. They are typically visual, but can sometimes contain other types of disruptions. Each symptom often develops gradually over several minutes and might last up to 60 minutes.

Visual disturbances such as seeing varied shapes, bright spots or flashes of light, vision loss, pins and needles feelings in an arm or leg, weakness or numbness in the face or one side of the body, and difficulty speaking are examples of migraine auras.

If left untreated, a migraine can continue anywhere from 4 to 72 hours. The frequency with which migraines occur varies from person to person. Migraines can occur infrequently or multiple times per month. During a migraine, you may have the following symptoms: pain, generally on one side of your head but sometimes on both, pain that throbs or pulses, sensitivity to light, sound, and sometimes smell and touch, nausea, and vomiting. You may feel fatigued, bewildered, and washed out for up to a day after a migraine attack. Some people claim to be elated. The pain may return briefly as a result of the rapid head movement.

Extract B : Home Remedies for Migraine

We've all had headaches, but if this one is the worst, it's most likely a migraine attack. Migraine symptoms begin slowly and steadily worsen, leaving you with a headache on one side of your head. Many people believe that a migraine is only a headache but, in fact, it can cause a variety of other symptoms. Did you know that migraine affects over half of the adult population, and that women are three times as likely than men to suffer from it? Fortunately, there are numerous migraine treatments available, just as there are for any other disease.

You may be able to control migraine symptoms at home with herbs and lifestyle modifications. However, these herbs and treatments should not be used in place of modern medicine. If home remedies do not improve your symptoms, consult your doctor. These are some common kitchen products that can be used as migraine cures at home.

In a 2014 clinical trial, Mehdi et al. discovered that ginger was useful in preventing and controlling migraine symptoms. Ginger includes several phytochemicals that may be helpful for headache relief. To alleviate migraine symptoms, make some ginger tea. Boil some freshly crushed ginger rhizome in water to make ginger tea. Allow it to simmer for a few minutes before straining and serving your ginger tea.

Caffeine is abundant in herbs such as coffee and guarana berries. Caffeine has the potential to both ease and trigger migraine symptoms. Caffeine should be used with caution. Caffeine in tiny amounts may be the best option for you.

Chamomile oil aromatherapy may help with migraine symptoms. Chamomile may also relieve nausea, which is a typical migraine symptom. To use chamomile to treat migraines, heat some chamomile oil and inhale the vapors through a vaporizer. You can also prepare chamomile tea by steeping chamomile tea leaves in hot water. Pour it into a cup. To enhance the flavor, honey can be added. And your chamomile tea is finished.

Peppermint oil, vapors, and extract have all been shown to help with migraines. Peppermint may help ease migraine symptoms. To utilize peppermint oil, heat some oil and use a vaporizer to inhale the fumes. Boiling peppermint leaves in water yields peppermint tea. Allow the mixture to simmer for a while. Fill a cup halfway with peppermint tea.

Here are some lifestyle adjustments that can aid in the prevention and relief of migraine symptoms. Drink plenty of water to avoid dehydration, especially if you've vomited as a result of a migraine. Furthermore, drinking water may help manage the headache produced by dehydration. You can also try different fluids besides water, such as juices and soups. Keep away from any liquids that may cause your migraine.

To alleviate migraine symptoms, try resting in a calm and dark place. Furthermore, while resting, you may want to catch up on some sleep. This will assist you in coping with migraine symptoms and waking up rejuvenated and calm. Trying certain relaxation techniques may aid with stress reduction. Relaxing treatments such as yoga, massage, conscious breathing, meditation, and exercise can be tried. You might also put a damp towel or cool cloth over your head.

Although some research demonstrate that the given herb and home cures for migraine are beneficial, they are insufficient. Large-scale human study is required to determine the extent to which these home cures help human health. They should be used with caution and never as a substitute for medical therapy.

Question: 8 of 30

Both extracts discuss on ..

- **A.** Abdominal pain
- **B.** Severe headache
- **C.** Brain tumor
- **D.** Fashion

Question: 9 of 30

What is an aura in relation to a migraine?

- **A.** A medicine that can prevent or lessen migraines
- **B.** A severe throbbing pain or pulsing sensation in the head
- **C.** A warning symptom that can occur before or during migraines
- **D.** None of the above

Question: 10 of 30

What might make the pain of a migraine return even after the initial attack has ended?

- **A.** Sudden head movements
- **B.** Eating spicy food
- **C.** Drinking caffeine
- **D.** None of the above

Question: 11 of 30

How does ginger help with migraines?

- **A.** Ginger can relieve dehydration caused by migraines.
- **B.** Ginger contains phytochemicals that can manage headaches.
- **C.** Ginger can be made into a vaporizer for inhalation.
- **D.** Ginger has been proven to be an alternative to modern medicine.

Question: 12 of 30

Why should home remedies for migraines be used cautiously?

- **A.** Home remedies are not effective in managing migraine symptoms.
- **B.** Home remedies can cause more harm than good.
- **C.** The benefits of home remedies have not been scientifically proven.
- **D.** Home remedies should only be used as a substitute for modern medicine.

Direction for Questions: 13 to 15

Read the extracts and then answer the questions.

Extract A:
Cape Town is a port city on South Africa's southwest coast, on a peninsula beneath the imposing Table Mountain. Slowly rotating cable cars climb to the mountain's flat top, from which there are sweeping views of the city, the busy harbor and boats heading for Robben Island, the notorious prison that once held Nelson Mandela, which is now a living museum. Cape Town, a vibrant city in South Africa, is famous for its breathtaking natural beauty, including Table Mountain and stunning beaches. It has a rich history, diverse culture, and delicious cuisine. However, Cape Town also faces challenges like economic inequality and water scarcity. With its lively arts scene and welcoming atmosphere, Cape Town offers a unique and unforgettable experience for visitors.

Extract B:
Canberra is the capital city of Australia. Founded following the federation of the colonies of Australia as the seat of government for the new nation, it is Australia's largest inland city and the eighth-largest Australian city overall. Canberra, the capital city of Australia, is known for its planned layout and beautiful architecture. It serves as the political center and houses important institutions like Parliament House and the Australian War Memorial. With its clean, green environment and high standard of living, Canberra is a great place to live and work.

Extract C:
Ontario is a home to Ottawa, Canada's capital, known for Parliament Hill's Victorian architecture and the National Gallery, featuring Canadian and indigenous art. Toronto, Ontario's capital, is home to the 553m-high CN Tower, with expansive views from its revolving restaurant, as well as High Park, site of a rare oak savannah habitat. Ontario, a province in Canada, is known for its stunning landscapes, including the iconic Niagara Falls and Algonquin Provincial Park. With its diverse population and multicultural cities like Toronto and Ottawa, Ontario offers a vibrant and inclusive society. It is also a hub for innovation and education, with renowned universities and thriving industries.

Extract D:
Sydney, capital of New South Wales and one of Australia's largest cities, is best known for its harbor front Sydney Opera House, with a distinctive sail-like design. Massive Darling Harbor and the smaller Circular Quay port are hubs of waterside life, with the arched Harbor Bridge and esteemed Royal Botanic Garden nearby. Sydney Tower's outdoor platform, the Skywalk, offers 360-degree views of the city and suburbs. Sydney, the largest city in Australia, is famous for its iconic landmarks like the Sydney Opera House and Sydney Harbor Bridge. With its stunning beaches, vibrant food scene, and lively arts and culture, Sydney is a city that never fails to captivate. It is also a major economic and financial center, attracting people from all over the world. Sydney truly offers a unique blend of natural beauty, urban sophistication, and laid-back lifestyle.

Question: 13 of 30

Which of the extracts refer to the harbor front Sydney opera house?

- **A.** Extract A
- **B.** Extract B
- **C.** Extract C
- **D.** Extract D

Question: 14 of 30

Which of the extracts describe context of stunning beaches, vibrant food scene, and culture?

- **A.** Extract A
- **B.** Extract B
- **C.** Extract C
- **D.** Extract D

Question: 15 of 30

Which of the extracts mention Robben island?

- **A.** Extract A
- **B.** Extract B
- **C.** Extract C
- **D.** Extract D

Direction for Questions: 16 to 22

Seven sentences have been removed from the text. Choose from the sentences (A – H) the one which fits each gap (1 – 7). There is one extra sentence which you do not need to use.

The Doctor-Patient Relationship: Trust, Compassion, and Collaboration

The doctor-patient relationship is a sacred bond built on trust, compassion, and collaboration.**1....** This relationship goes beyond the mere exchange of medical knowledge and treatment; it encompasses empathy, understanding, and shared decision-making. In this essay, we will explore the importance of the doctor-patient relationship and its impact on healthcare outcomes.

2.... Patients must trust their doctors to provide them with the best possible care, and doctors must trust their patients to be honest and forthcoming about their symptoms and medical history. Trust is the bedrock upon which effective communication and collaboration are built.**3....** Trust also empowers patients to actively participate in their healthcare decisions, knowing that their doctor has their best interests at heart.

Compassion is another essential element of the doctor-patient relationship. Doctors must approach their patients with empathy, recognizing the vulnerability and anxiety that often accompany illness or injury. **4...** It is through compassion that doctors can truly understand the physical, emotional, and psychological impact of a patient's condition, leading to more effective and holistic care.

Collaboration is a key aspect of the doctor-patient relationship. Gone are the days of the paternalistic approach, where doctors made decisions on behalf of their patients without considering their input. **5...** Doctors must provide patients with the necessary information about their condition, treatment options, and potential risks and benefits. Patients, on the other hand, must actively engage in their care, ask questions, and voice their preferences. Through collaboration, doctors and patients can develop personalized treatment plans that align with the patient's goals and values.

6.... A strong doctor-patient relationship leads to improved patient satisfaction, adherence to treatment plans, and health outcomes. It reduces medical errors and unnecessary tests or procedures. Moreover, it creates a positive healthcare experience that instills confidence and trust in the healthcare system.

7.... It is a partnership that recognizes the importance of empathy, effective communication, and shared decision-making. This relationship has a profound impact on healthcare outcomes, ensuring that patients receive the best possible care and experience. By nurturing and fostering the doctor-patient relationship, we can create a healthcare system that is patient-centered, compassionate, and effective.

A. ***At the core of the doctor-patient relationship is trust.***
B. ***It allows patients to feel comfortable sharing their concerns and fears, enabling doctors to make accurate diagnoses and develop appropriate treatment plans.***
C. ***Today, healthcare is a partnership, with doctors and patients working together to achieve the best possible outcomes.***
D. ***The doctor-patient relationship not only impacts individual patients but also has broader implications for healthcare as a whole.***

E. ***Compassion involves actively listening to patients, acknowledging their emotions, and treating them with dignity and respect.***
F. ***By nurturing and fostering the doctor-patient relationship, we can create a healthcare system that is patient-centered, compassionate, and effective.***
G. ***It is a unique partnership that forms the foundation of effective healthcare delivery.***
H. ***In conclusion, the doctor-patient relationship is a sacred bond based on trust, compassion, and collaboration.***

Question: 16 of 30

Which one fits in (1) ?

- **A.** Sentence A
- **B.** Sentence B
- **C.** Sentence G
- **D.** Sentence H

Question: 17 of 30

Which one fits in (2) ?

- **A.** Sentence A
- **B.** Sentence B
- **C.** Sentence C
- **D.** Sentence D

Question: 18 of 30

Which one fits in (3) ?

- **A.** Sentence A
- **B.** Sentence D
- **C.** Sentence B
- **D.** Sentence E

Question: 19 of 30

Which one fits in (4) ?

- **A.** Sentence A
- **B.** Sentence B
- **C.** Sentence H
- **D.** Sentence E

Question: 20 of 30

Which one fits in (5) ?

- **A.** Sentence A
- **B.** Sentence B
- **C.** Sentence C
- **D.** Sentence D

Question: 21 of 30

Which one fits in (6) ?

- **A.** Sentence A
- **B.** Sentence B
- **C.** Sentence C
- **D.** Sentence D

Question: 22 of 30

Which one fits in (7) ?

- **A.** Sentence A
- **B.** Sentence G
- **C.** Sentence H
- **D.** Sentence C

Direction for Questions: 23 to 27

Read the poem below then answer the questions.

Only A Dad
***By* Edgar A. Guest**

Only a dad, with a tired face,
Coming home from the daily race,
Bringing little of gold or fame,
To show how well he has played the game,
But glad in his heart that his own rejoice
To see him come, and to hear his voice.

Only a dad, with a brood of four,
One of ten million men or more.
Plodding along in the daily strife,
Bearing the whips and the scorns of life,
With never a whimper of pain or hate,
For the sake of those who at home await.

Only a dad, neither rich nor proud,
Merely one of the surging crowd
Toiling, striving from day to day,
Facing whatever may come his way,
Silent, whenever the harsh condemn,
And bearing it all for the love of them.

Only a dad, but he gives his all
To smooth the way for his children small,
Doing, with courage stern and grim,
The deeds that his father did for him.
This is the line that for him I pen,
Only a dad, but the best of men.

Question: 23 of 30

In the poem, what does the phrase "the daily race" (line 2) most likely refer to?

- **A.** A marathon
- **B.** A competitive sport
- **C.** The challenges of daily life and work
- **D.** A horse race

Question: 24 of 30

What is the deeper meaning conveyed by the line "But glad in his heart that his own rejoice" ?

- **A.** The father finds joy in his children's happiness.
- **B.** The father is happy because he has achieved fame.
- **C.** The father is glad to be done with his daily work.
- **D.** The father is unhappy despite his children's joy.

Question: 25 of 30

What does the poet mean by "The deeds that his father did for him" ?

- **A.** The father's actions and sacrifices for his children
- **B.** The father's achievements and successes
- **C.** The father's hobbies and interests
- **D.** The father's mistakes and failures

Question: 26 of 30

What is the speaker's attitude towards the father in the poem?

- **A.** Admiration
- **B.** Indifference
- **C.** Criticism
- **D.** Envy

Question: 27 of 30

What is the central message or theme of the poem "Only a Dad"?

- **A.** Fathers are unimportant in the lives of their children.
- **B.** The sacrifices of fathers often go unnoticed.
- **C.** Fathers should focus on achieving fame and wealth.
- **D.** Children should express their gratitude to their fathers regularly.

Direction for Questions: 28 to 30

Read the poem below then answer the questions.

Nothing Gold Can Stay
***By* Robert Frost**

Nature's first green is gold,
Her hardest hue to hold.
Her early leaf's a flower;
But only so an hour.
Then leaf subsides to leaf.
So Eden sank to grief,
So dawn goes down to day.
Nothing gold can stay.

Question: 28 of 30

Which line in the poem uses metaphor to describe nature's initial state?

- **A.** "Nature's first green is gold"
- **B.** "Her hardest hue to hold"
- **C.** "So dawn goes down to day"
- **D.** "Nothing gold can stay"

Question: 29 of 30

How does the poet use sound alliteration in the poem?

- A. "But only so an hour."
- B. "So Eden sank to grief,"
- C. "Nothing gold can stay."
- D. "Her hardest hue to hold."

Question: 30 of 30

What deeper meaning or message does the poem convey about life?

- A. Life is eternal and unchanging.
- B. Life is filled with constant growth and renewal.
- C. Life is brief and transient, and beauty fades.
- D. Life is unpredictable and chaotic.

Preptive Prepare Better

Series B Mathematical Reasoning Trial Test

Name: ______________________________

Date: ______________________________

Question: 1 of 35

How many possible combinations of 1 pencil and 1 rubber cap can be formed from 3 pencils and 4 rubber caps?

- **A.** 7
- **B.** 10
- **C.** 12
- **D.** 14

Question: 2 of 35

Find the hidden numbers to get the same value when the numbers in the boxes below are added across and down.

?	4	6
?	6	5
?	8	7

- **A.** 2, 3, 6
- **B.** 8, 7, 3
- **C.** 6, 2, 3
- **D.** 7, 6, 2

Question: 3 of 35

According to the pattern, what should be the numbers for A and B respectively?

2	3	4	24
1	A	3	15
4	2	B	32
8	30	48	

- **A.** 6 and 3
- **B.** 3 and 3
- **C.** 5 and 4
- **D.** 4 and 3

Question: 4 of 35

Ricky paid $473 for 5 similar cushions and 3 similar blankets. Shawn paid $339 for 3 similar cushions and 3 similar blankets.
How much is the cost of 3 similar cushions?

- **A.** 201
- **B.** 302
- **C.** 134
- **D.** 339

Question: 5 of 35

Namo saved $30 more than Chitra. Aria saved 3 times as much money as Namo. The 3 girls saved a total of $300. How much money did Namo save?

- **A.** 22
- **B.** 23
- **C.** 33
- **D.** 66

Question: 6 of 35

Aston has $180 more than Bob. Kevin has twice as much as Aston. The three children have a total amount of $1140. How much does Bob have?

- **A.** 110
- **B.** 120
- **C.** 140
- **D.** 150

Question: 7 of 35

There were 38 children at a fun-fair. Every boy received 2 balloons each and every girl received 4 balloons each at the fun-fair. A total of 126 balloons were given to the children.
How many boys were there at the fun-fair?

- **A.** 10
- **B.** 11
- **C.** 12
- **D.** 13

Question: 8 of 35

Amirah had a collection of 20 old coins. The value of the coins was $6.40 in total.
In her collection, there were only 20-cent coins and 50-cent coins.
How many 50-cent coins did she have?

- **A.** 6
- **B.** 7
- **C.** 9
- **D.** 8

Question: 9 of 35

Lynn had 93 pencils and pens. After giving away $\frac{3}{4}$ of her pens and 28 pencils, she had an equal number of pencils and pens left.
How many pencils did she have at first?

- A. 65
- B. 42
- C. 41
- D. 40

Question: 10 of 35

Lucas had some sweets.
If he gave 7 sweets to each of his friends, he would be short of 4 sweets. If he gave 6 sweets to each of them, he would have 3 sweets left.

How many sweets did Lucas have?

- A. 14
- B. 45
- C. 17
- D. 7

Question: 11 of 35

Jiemin has some orange juice. If he fills 20 identical cups with the orange juice, he will have 600 ml of the orange juice left. If he tries to fill 6 identical jugs with the orange juice instead, he will be short of 400 ml of the orange juice. The capacity of a jug is the same as that of 4 such cups.
What is the capacity of a jug?

- A. 1500ml
- B. 2000ml
- C. 5000ml
- D. 1000ml

Question: 12 of 35

A drink stall owner sold $\frac{1}{9}$ of his canned drinks in the morning and $\frac{3}{4}$ of the remaining canned drinks in the afternoon. After he received a delivery of another 200 canned drinks, he found that he had 32 canned drinks more than what he had at first. How many canned drinks did the stall owner have at first?

- A. 216
- B. 215
- C. 168
- D. 24

Question: 13 of 35

A fruit stall sells apples, oranges and pears. The ratio of the number of pears to the total number of fruits is 2 : 5. There are 300 apples and 180 oranges.

Find the number of pears at the stall.

- **A.** 360 pears
- **B.** 160 pears
- **C.** 320 pears
- **D.** 480 pears

Question: 14 of 35

The bar chart below shows the number of sales of 4 types of animals at a pet store.

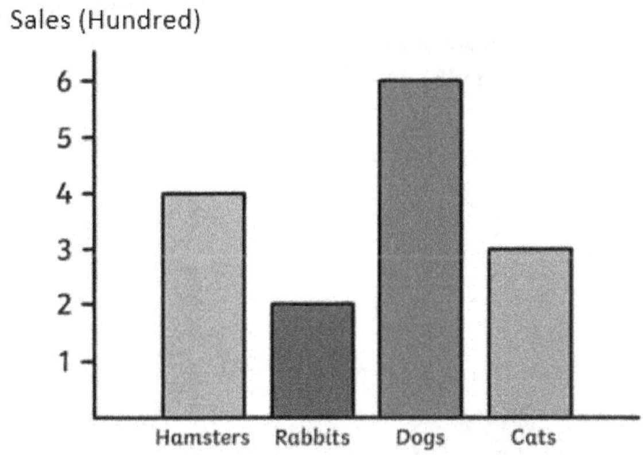

Which of the following is a correct statement?

- **A.** The sales of hamsters are the highest.
- **B.** The difference in sales between dogs and cats is 300.
- **C.** The sum of sales of rabbits and hamster is 500.
- **D.** None of the above.

Question: 15 of 35

Diagram below shows a triangle

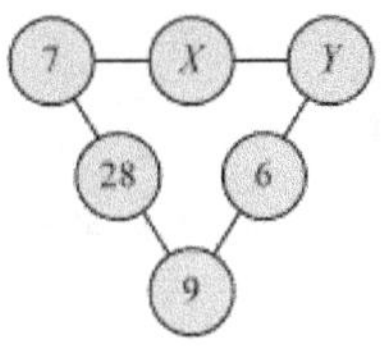

The total of the 3 numbers on each side of the triangle is the same. What are the values of X and Y?

- **A.** X = 8, Y = 39
- **B.** X = 7, Y = 29
- **C.** X = 8, Y = 29
- **D.** X = 10, Y = 29

Question: 16 of 35

The graph shows the number of ice creams sold.

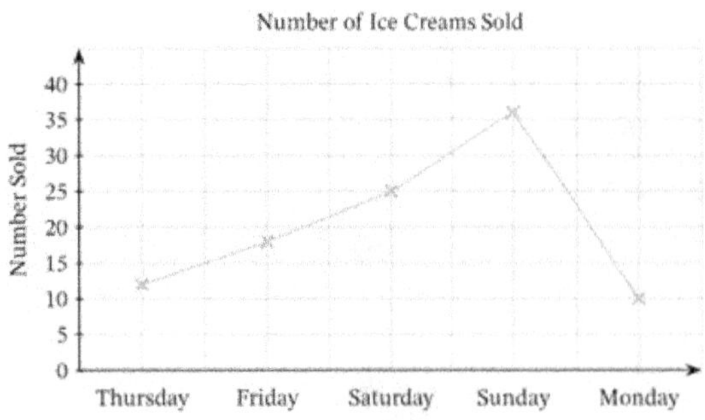

Which of the following is correct?

A. The highest sales are on Saturday

B. The number of ice creams sold on Saturday and Monday is 35.

C. The lowest sales are on Sunday.

D. None of the above.

Question: 17 of 35

The graph below shows the temperature on 1st September 2022.

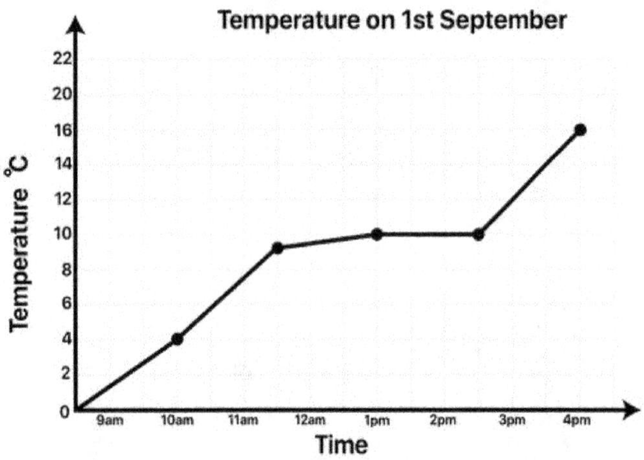

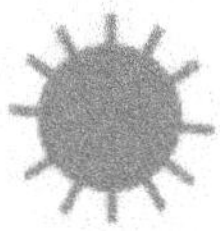

Which of the followings is a wrong statement based on the graph?

A. It is 16°C at 4 in the evening

B. It is 10°C at 1 in the afternoon

C. It is 4°C at 11 in the morning

D. The highest temperature recorded is at 4 pm

Question: 18 of 35

The table shows the number of beads in a container.

Color	Number Of Beads
White	50
Red	30
Blue	$\frac{2}{5}$ of the number of white beads

Daisy uses $^3/_5$ white beads and 40% of red beads to embroider her dress.
Count the number of beads left in the container.

A. 25
B. 84
C. 58
D. 75

Question: 19 of 35

Below shows a dialogue between four students who collect donations.

Larry : I was able to collect $2.50.
Mei Ling : The amount of money I collected was 90 cents more than Larry.
Raj : My collection money is the same as Mei Ling's money.
John : If added to the money I collected, the total is $12.

Based on the dialogue, count the money that John successfully collected.

A. $2.70
B. $3.70
C. $4.50
D. $2.30

Question: 20 of 35

Rayyan sets an alarm at 5.40 a.m. When the alarm rings, he pressed the snooze button and goes back to sleep. The alarm rang every 5 minutes. If Rayyan wakes up at 6.10 a.m. when alarm rang, how many times does the alarm ring?

A. 5
B. 6
C. 10
D. 7

Question: 21 of 35

At a party, each girl was given 3 balloons and each boy was given 5 balloons. There were 32 children at the party and a total of 116 balloons were given out. How many boys were there at the party?

A. 10
B. 11
C. 3
D. 8

Question: 22 of 35

Figure A is made up of a square and 8 identical triangles. Each triangle has a height of 7 cm and a base of 3.5 cm. Find the area of Figure A.

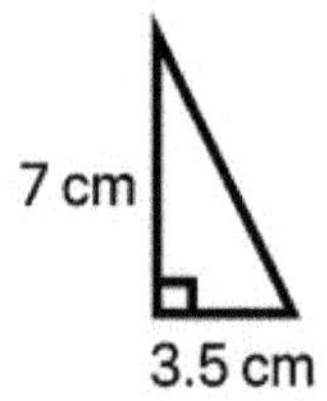

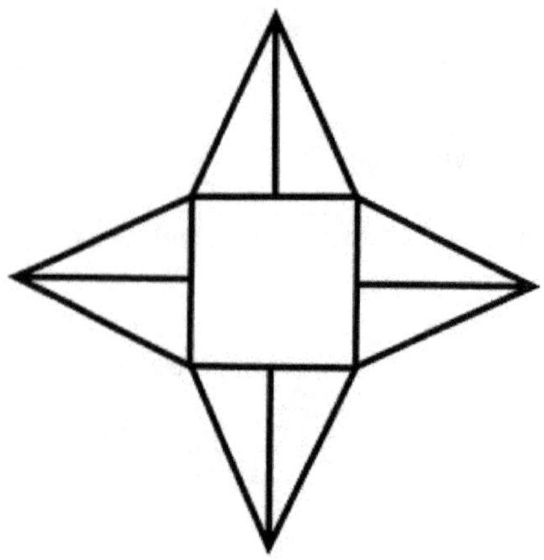

Figure A

- A. 124 cm^2
- B. 139 cm^2
- C. 149 cm^2
- D. 147 cm^2

Question: 23 of 35

If the length of the side of a square is doubled, what is the ratio of the areas of the original square to the area of the new square?

- A. 1:5
- B. 1:4
- C. 1:3
- D. 1:9

Question: 24 of 35

In the figures given below, ABCD is a rhombus.

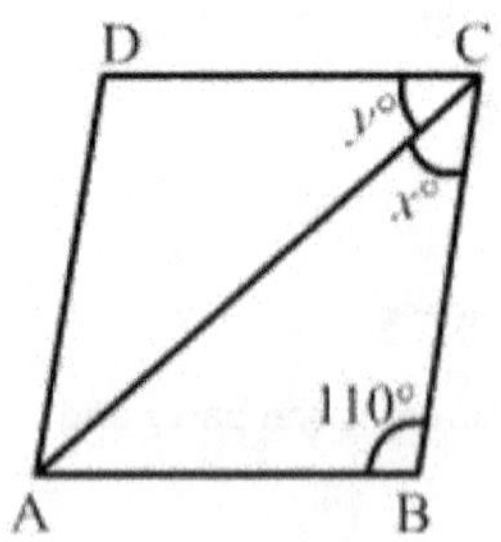

Find the value of x and y.

- A. x = 35°, y=35°
- B. x = 45°, y = 35°
- C. x = 55°, y = 45°
- D. x = 35°, y = 65°

Question: 25 of 35

The ratio of the number of boys and girls in a college is 7: 8. If the percentage increase in the number of boys and girls is 20% and 10% respectively, what will be the new ratio?

- **A.** 8:9
- **B.** 17:18
- **C.** 21:22
- **D.** Cannot be determined

Question: 26 of 35

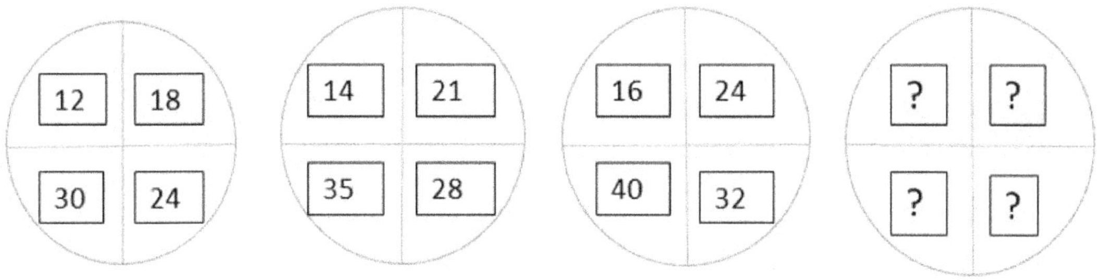

What are the 4 numbers that are suitable for the quarters of the 4 circles?

- **A.** 22,28,34,45
- **B.** 18,27,36,45
- **C.** 24,32,42,54
- **D.** 26,35,40,58

Question: 27 of 35

There are white keys in a small key board that are named as C, D, E, F, G, A and B. What is the probability of pressing E note by a blind folded person in his all 2 chances?

- **A.** $^{1}/_{7}$
- **B.** $^{2}/_{7}$
- **C.** $^{1}/_{49}$
- **D.** $^{2}/_{49}$

Question: 28 of 35

Simplify and find the value in scientific notation.
$6.78 \times 10^8 + 17.5 \times 10^7$

- **A.** 24.28×10^{15}
- **B.** 24.28×10^{7}
- **C.** 8.53×10^{16}
- **D.** 8.53×10^{8}

Question: 29 of 35

Which of the below options is correct?

A box contains 9 sprite bottles, 18 Coca-Cola bottles and 27 Fanta bottles.

The ratio 1:6 compares to ...

- **A.** None of the options
- **B.** Sprite bottle to Fanta bottles
- **C.** Fanta bottles to Coca-Cola bottles
- **D.** Sprite bottles to all the bottles

Question: 30 of 35

Aina is a worker in a shoe shop. She has 5 types of shoes to keep in racks. How many different ways can she arrange these shoes in a row?

- **A.** 5
- **B.** 25
- **C.** 60
- **D.** 120

Question: 31 of 35

John had 4 times as many stickers as Mark. Dan had 2 times as many stickers as John. Mark had 420 stickers fewer than John. How many stickers did Dan have?

- **A.** 1440
- **B.** 1560
- **C.** 1120
- **D.** 1020

Question: 32 of 35

Four children shared some sweets among themselves. Adam took $^1/_8$ of the total number of sweets while Barry took $^1/_4$ of them. Kaden took $^1/_4$ of the remaining sweets and Shaun took the rest.
If Barry took 20 more sweets than Adam, how many sweets did Shaun take?

- **A.** 75
- **B.** 160
- **C.** 25
- **D.** 100

Question: 33 of 35

A train carrying 72 adults left Stop X for Stop Y.
$^6/_{11}$ of the passengers were adults and the rest were children.
At Stop Y, $^1/_4$ of the children got off the train.

How many children were on the train when the train left Stop Y?

- **A.** 25
- **B.** 15
- **C.** 45
- **D.** 60

Question: 34 of 35

The table below shows the different types of fruits Mr. Malhotra sold.

Fruits	Number of fruits sold
Oranges	50
Pears	30
Mangoes	
Avocadoes	20

The number of mangoes sold was $^1/_5$ of the total number of fruits sold. How many mangoes were sold?

A. 30

B. 35

C. 20

D. 25

Question: 35 of 35

A box of milk has 2.5 L milk. Jenny drank half of the milk. Lisa drank 20% of the milk left. Rosie put another 190ml of milk into the box.
How much milk is in the box now?

A. 1,190ml

B. 1,458ml

C. 1,100ml

D. 1,300ml

Preptive Prepare Better

Series B Thinking Skill Trial Test

Name: ______________________________

Date: ______________________________

Question: 1 of 40

Read the statements and decide the validity of conclusion.
Note : For the purpose of this question, statements are always to be taken as correct, even if they contradict commonly known facts.

Statement I : Pictures can tell a story.
Statement II: All storybooks have pictures.
Statement III: Some storybooks have words.

Conclusion: Some storybooks have both words and pictures.

- **A.** Conclusion is always true.
- **B.** Conclusion is always false.
- **C.** Conclusion is sometimes true.
- **D.** Can't decide.

Question: 2 of 40

Gloriosa superba, commonly known as the flame lily, is species of flowering plant. It is a climbing lily species that is equally beautiful and poisonous. Every part of the plant is poisonous and it's toxic enough to cause human and animal fatalities if ingested. It has been used to commit murder, to achieve suicide, and to kill animals. This flower has **very unique** flame-like six tepals generally bright red to orange, with yellowish bases.

Niko: "A person has died after eating a random flower in a forest as the flower was looking beautiful. That flower must be a flame lily."

Soya: "Yesterday I saw there is a plant with flowers which has flame-like six tepals generally bright red to orange, with yellowish bases, so that must be a Gloriosa superba plant"

If the given information is true, whose reasoning must be correct?

- **A.** Niko only
- **B.** Soya only
- **C.** Both Niko nor Soya
- **D.** Neither Niko nor Soya

Question: 3 of 40

Robin loves dogs and rabbits but she hates cats. James loves pets and he has two dogs, a cat, and a rabbit. He loves them all. Ron doesn't like dogs, but he has a cat and two rabbits.
Which of the following is correct about the above situation?

- **A.** Robin hates all the pets of James
- **B.** Robin loves all the pets of Ron
- **C.** Ron hates all the pets of Robin
- **D.** James loves all of Robin's pets

Question: 4 of 40

You are given an A4 paper. If you fold it as shown in the dotted line in figure 1 and then cut along the dotted line shown in figure 2, which are the resulting shapes?

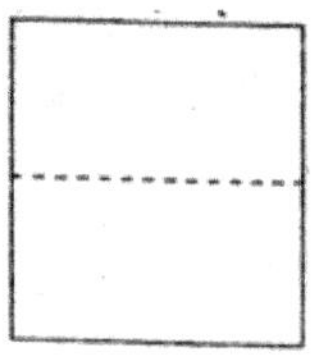

Figure 1

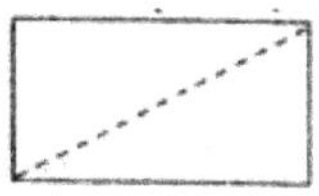

Figure 2

- **A.** Four triangles
- **B.** Two triangles
- **C.** Three triangles
- **D.** A rectangle and two triangles

Question: 5 of 40

Neville: "If you win the chess championship, you can beat Ron."

Jonathan: "I won a chess match with Ron. So, I can win the championship."

If Neville is correct, Jonathan is ...

- **A.** Correct
- **B.** Wrong
- **C.** Both correct and wrong
- **D.** Cannot decide

Question: 6 of 40

A: "There is at least one student who got full marks for the assessment. "

B: "Everyone got at least 1 mark."

Which of the following is correct?

- **A.** A and B have the same meaning
- **B.** A has the opposite meaning of B
- **C.** B is the opposite of A
- **D.** None of the above

Question: 7 of 40

If someone is good at running, he is also good at long jump.
Athletes who are good at long jump are tall.

Which of the following is a possible conclusion based on above two statements ?

- **A.** Tall people are good at running
- **B.** Tall people are good at the long jump
- **C.** Good runners are tall
- **D.** None of the above

Question: 8 of 40

In the school music band none of the member's height is less than 168cm.
All the members of the school band must know how to play the trumpet properly.

Darren: "Susan plays the trumpet very well also he has a height of 175cm. He must be a member in the school music band"

Noah: "Paul is willing to join the school music band but he is only 164cm. Definitely, he can't join the band"

If the information in the box is true, whose reasoning must be correct?

- **A.** Darren only
- **B.** Noah only
- **C.** Both Darren and Noah
- **D.** Neither Darren nor Noah

Question: 9 of 40

A microorganism is a living organism which is too small to be seen with the eye but visible under a microscope. Viruses are even smaller therefore they can only be seen through an electron microscope. The term microorganisms do not include viruses, which are generally classified as non-living as they have no energy metabolism.

Paul: "If there is an organism that doesn't have an energy metabolism, then that must be a virus."
Kia: "As viruses are not microorganisms, in some situations they might be seen with the eye"

If the given information is true, whose reasoning must be true?

- **A.** Paul only
- **B.** Kia only
- **C.** Both Paul and Kia
- **D.** Neither Paul nor Kia

Question: 10 of 40

If Mr. Lucas gets the majority of the votes in the election, he will become a parliament member.
If he becames a parliament member he will definitely speak for the low-income people who suffer from taxes, otherwise he will not speak.
If he speaks for the low-income people who are suffering from taxes the government might consider cutting some taxes.

If the given information is true, which one of the following must be true?

A. Mr. Lucas got the majority of the votes in the election but he didn't speak for the low-income people who suffer from taxes

B. Mr. Lucas did not get the majority of the votes in the election but he spoke for the low-income people who suffer from taxes.

C. Mr. Lucas got the majority of the votes in the election and he spoke for the low-income people who suffer from taxes.

D. Mr. Lucas got the majority of the votes in the election and the government cut some taxes.

Question: 11 of 40

Lithium is a soft, silvery-white, alkali metal. Lithium metal reacts vigorously with water. Of the alkali metals, ***only lithium reacts with nitrogen****, and it forms a nitride.*

Charles: "If an alkali metal is soft, silvery-white and reacts vigorously with water, then that must be Lithium"

Karen: "If there is an alkali metal that reacts with nitrogen and forms its nitride that must be Lithium"

If the information is true, whose reasoning must be correct?

A. Charles only

B. Karen only

C. Both Charles and Karen

D. Neither Charles nor Karen

Question: 12 of 40

Below is an excerpt from an article.

Advances in technology have contributed to the meticulous use of robotics. The most important benefit of Robots is they are used to do repetitive actions or jobs considered too dangerous for humans. Robots are ideal for doing very risky jobs like nuclear cleaning, firefighting, searching and rescuing in disasters etc.

Fredric: "I don't agree with what the article prioritises. Robotic surgeries, allows doctors to perform many types of complex procedures with more precision, flexibility and control than is possible with conventional techniques. Robotic surgeries has reduced risk of infection or complications. There are a lot of benefits in a robotic surgeries than a regular surgery like less scarring because of smaller incisions and fewer sutures, briefer hospital stays and faster recovery, more rapid return to normal activities."

Which one of the following statements, if true, describes why Fredric disagrees with the article?

A. Robotics has been popular since before the development of technology.

B. Robotic surgery doesn't offer patient benefits when compared to traditional open surgery.

C. Robots cannot be used to do repetitive actions or jobs considered too dangerous for humans.

D. Robotics has its most valuable achievements in the surgical field.

Question: 13 of 40

Following table contains the birthdays of 4 students.

Name	Birthday
Kevin	1998.05.20
Ross	1999.02.25
Helen	1999.01.20
Lily	1998.05.21

Who is the youngest?

- **A.** Kevin
- **B.** Helen
- **C.** Lily
- **D.** Ross

Question: 14 of 40

Sam : "All the architects are good at drawing arts."
Ken : "Lily got full marks for art."

Which of the below is true according to **both Sam and Ken** ?

- **A.** Lily is an architect
- **B.** Lily can be an architect
- **C.** Lily is going to be an architect one day
- **D.** No conclusion

Question: 15 of 40

Julie, Michael and Rachel are discussing about current affairs.

Julie: "Every businessman owns more than one vehicle."
Michael: "If someone has many vehicles the maintenance cost is higher."
Rachel: "I do know that Ross is rich."

If all three are correct which of the following is a possible conclusion **about Ross** ?

- **A.** Ross has one vehicle
- **B.** Ross pays a lot for vehicle maintenance
- **C.** Ross is a good friend of Rachel
- **D.** None of the above

Question: 16 of 40

Cantharidin are toxic agent released by blister beetles which can cause irritation or blistering when it comes into contact with the skin, mouth, eyes, throat or even in the digestive tract when eaten. Cantharidin poisoning is most common in horses who consume contaminated alfalfa hay. Horses can die as a result of exposure. Cantharidin poisoning affects other species as well, including dogs, cats, rabbits, and humans.

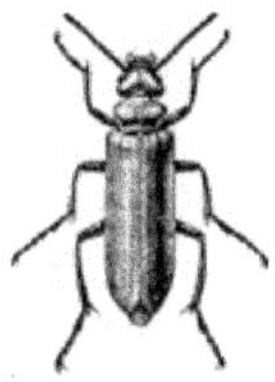

Which of the following must also be true?

A. Skin irritation can not occur when cantharidin came into contact with the skin.

B. Alfalfa hay with blister beetle should be fed to horses.

C. Cats who eat a blister beetle can also suffer from cantharidin poisoning and it can be fatal.

D. Alfalfa hay which is suspected to not be contaminated must not be fed to horses.

Question: 17 of 40

Rose either drink coffee or cold water every day without fail. If it is cold and raining then Rose drinks coffee. If it is sunny and not raining then Rose drinks cold water. Sometimes it can be sunny and cold.

Which of the following statement must be true?

A. If it is raining and not sunny then Rose drinks her coffee.

B. If it is raining and Rose does not drink her coffee then it is not cold.

C. If it is not sunny and it is cold then Rose drinks her coffee.

D. If it is sunny and not raining then Rose drinks her coffee.

Question: 18 of 40

A terpene called camphor is frequently found in creams, ointments, and lotions. The oil produced by steam distillation after the camphor tree's wood is removed is known as camphor oil. To ease pain, irritability, and itching, use camphor topically. Camphor can induce convulsions, difficulty breathing, and even death when taken orally. High quantities of camphor can also be dangerous when applied topically or breathed. They may result in skin rashes or convulsions.

If the information is true, which of the following is true?

- **A.** Camphor cannot cause skin irritation and must not be used.
- **B.** Camphor can be mixed with a hot water and taken orally to relieve sore throat.
- **C.** Camphor must be used with caution and in small appropriate amounts.
- **D.** Camphor can be mixed with tea and coffee to relieve stomach pain.

Question: 19 of 40

In the era of digital art tools, traditional painting has lost its significance.

Which of these statements weakens the above claim?

- **A.** Not all digital art tools offer the same level of creative freedom and depth as traditional painting.
- **B.** Only professional artists can truly appreciate the value of traditional painting techniques.
- **C.** Traditional painting has always been limited in terms of colour choices and texture effects.
- **D.** Digital art tools have completely replaced traditional painting methods in modern art schools.

Question: 20 of 40

Company A is calculating its annual sales. Which statements will help solve this calculation?

i. **Company A sells a piece of tile at $20, and it sold 10,000 pieces that year.**
ii. **The company only sells a single product.**

- **A.** Statement i is enough
- **B.** Statement ii is enough
- **C.** Both statements are necessary
- **D.** Both statements aren't enough to calculate the sales

Question: 21 of 40

A, B, C, D, E, F and G are in a straight line but not necessarily in that order.

i. A is next to G
ii. G is fourth from C's right
iii. D is between B and F
iv. C is 3rd from B's left and also on one end

Who are the people adjacent to A?

A. B and C
B. C and G
C. G only
D. F only

Question: 22 of 40

For one to be the chair of a committee of 15 members, there is a need for the maximum number of votes. If 5 people are contesting for the post, which of the following reasoning is correct?

Elizabeth: If every candidate gets at least one vote, the maximum the winner can get is 11 votes.
Emily: A win is guaranteed if a candidate gets 8 votes.

A. Elizabeth
B. Emily
C. None of them
D. Both statements

Question: 23 of 40

A certain country gives various currency denominations different names. The names are alpha, beta, gamma, ray, and omega. If 5 alphas make a beta, 3 betas make a gamma, 4 gammas make a ray, and 5 rays make an omega, how many betas are in 2 omegas?

A. 60
B. 40
C. 120
D. 150

Question: 24 of 40

According to certain research, tall men have quite large feet.

If the statement is true, which of the following statements is true?

i. If your grandfather is taller than your father and wears a shoe size 10, then size 9 could be ideal for your father.
ii. If your grandfather and father are the same height, their shoe size can be the same.
iii. Whereas your father is shorter than your grandfather, his shoe size can be bigger than your grandfather's.

A. All the above
B. None of the above
C. i and ii
D. i and iii

Question: 25 of 40

Both Eva and Adam are good at Math. Only Eva is good at History. Adam is good at Science, which Eva isn't.

Based on the above information about Eva and Adam, which one of the following must be correct?

- **A.** Both of them are good at Math and Science.
- **B.** Only Adam is not good at History
- **C.** Eva is not good at anything that Adam good at
- **D.** Both of them are good at all the subjects

Question: 26 of 40

If a student's attendance is more than 75% the teacher will add additional 5 marks for the written test, but only if the student has scored lower than 65 marks for the written test. To pass the written test a student must score at least 35 marks.

Diana: "I already have 80% attendance. So I will definitely pass the written test even if I score 30 marks in my written test."

Samantha: "I also have 79% attendance. So if I get 95 marks in my written test, I will get 100 marks as my final results."

If the above information is true, whose reasoning must be true?

- **A.** Diana only
- **B.** Samantha only
- **C.** Both Diana and Samantha
- **D.** Neither Diana and Samantha

Question: 27 of 40

Analia has planned to go to a carnival and to take dinner at a nearby hotel.

If Analia goes to the carnival she will likely go for a ride on the roller coaster.
If she goes on a roller coaster ride she will feel nauseous and dizzy.
If she feels nauseous and dizzy she will go back home without enjoying her dinner.

If the information given above is true, which of the following is not possible?

- **A.** Analia didn't go on a roller coaster but she was nauseous and dizzy.
- **B.** Analia went on a roller coaster ride but she went to the hotel for dinner.
- **C.** Analia went back home without enjoying her dinner but she even did not go on the roller coaster.
- **D.** Analia went to the carnival but she did not go for a ride on the roller coaster.

Question: 28 of 40

Rosy and Julie attended a life lesson workshop. The instructor told : "The best way to escape from a problem is to solve it."
Instructor asked everyone about thier understanding on her statement.

Rosy: Your life will be dull if you don't face a problem.
Julie: To escape from problems, you should always have some solutions with you.

Who's explanation is more accurate to the meaning of instructor's statement ?

A. Rosy is accurateJulie is accurate

B. Julie is accurate

C. Both Julie and Rosy are accurate

D. Neither Julie nor Rosy is accurate

Question: 29 of 40

There are 5 trees in a garden. All the trees were planted for over the past 5 years. The 1st tree was planted in the 1st year and 2nd tree was in the second year and so on. The total heights of all the trees is 100 metres.

Madison: " All the trees may have different heights".
Addison: "All the trees will have same height."

Whose argument is correct?

A. Madison is correct

B. Addison is correct

C. Both are correct

D. Both are wrong

Question: 30 of 40

Statement I: James is a good director.
Statement II: All directors are intelligent.

Based on above two statements, which conclusion is true ?

A. All intelligent are directors

B. James is intelligent

C. Both of the above

D. None of the above.

Question: 31 of 40

Maya's college is giving scholarships to students for further studies in France.

If a student has at least some knowledge of French, then they only have to pass the aptitude test. If they have none, then they either needed to get an excellent score on the aptitude test or follow the French course that the college offers and pass the French test and then pass the aptitude test.

Maya is fluent in French but failed to get the scholarship. What must have been the reason?

A. She didn't pass the French test.

B. She didn't get an excellent score on the aptitude test.

C. She failed the both French test and aptitude test

D. She failed the aptitude test

Question: 32 of 40

If someone is a lawyer, then that person has good knowledge about the law and is good at arguing.

Which of the following is the opposite of the above statement?

A. Some lawyersare good at arguing.

B. Some lawyers don't have good knowledge about the law.

C. Some lawyers are not good at arguing or don't have good knowledge about the law.

D. Some lawyers are bad at arguing and have good knowledgeof the law.

Question: 33 of 40

In a survey of pets that elementary school students 6 were having dogs, 7 were having cats and 4 were having birds. Students had no other animals as pets.

James: 17 students must have participated in the survey.

If the information is true, which one of the following sentences shows the mistake James has made?

A. Not only dogs, cats and birds there are so many other animals that kept as pets

B. The number of those who participated in the survey might be less than 17

C. There must be more than 17 students who participated in the survey

D. The number of those who participated in the survey might be more than 17

Question: 34 of 40

If lightning strikes today people won't go outside and tomorrow trees will look so green.
It's not lightning today.

Which of the following is a possible conclusion?

A. People won't go outside today.

B. Trees will look so green tomorrow.

C. People won't go outside today, and trees will look so green tomorrow.

D. None of the above.

Question: 35 of 40

If Rachel is wearing new clothes, then she has bought some new clothes or she has borrowed some clothes from Monica.

Which of the following has the same meaning as the above statement?

A. If Rachel is wearing new clothes and she didn't borrow clothes from Monica.

B. If Rachel is wearing new clothes and she didn't buynew clothes.

C. If Rachel didn't borrow clothes from Monica and didn't buy some new clothes then she is not wearing new clothes.

D. If Rachel is not wearing new clothes, then she didn't borrow clothes from Monica.

Question: 36 of 40

Every bike is fast.
Jonathan bought an expensive bike.

So, we can conclude that ..

- **A.** Jonathan's bike is fast.
- **B.** Jonathan's bike isn't fast because it's expensive.
- **C.** None of the above.
- **D.** Cannot decide.

Question: 37 of 40

There is a 10-day music festival that will be held from 1st to 10th of this month. Ticket prices for the music festival are shown below.

- 1 day = $20
- 2 consecutive days = $30
- 5 consecutive days = $60

Paul wants to go to the music festival on the 1st, 2nd, 7th, 8th, 9th, and 10th. What is the minimum amount Paul can spend to go to the music festival on all these days?

- **A.** $100
- **B.** $90
- **C.** $110
- **D.** $70

Question: 38 of 40

Professor Lucas: "End semester exam follows mid semester exam. All the students who got more than 75 marks in the mid-semester exam will automatically get an A pass for the end semester exam, without taking end semester exam."

Helen: "Adam got 75 marks in the mid-semester exam. So, he will get an A pass for the end semester exam."

Kathy: "Chandler got 60 marks for the mid-semester exam. He may or may not get an A pass for the end semester exam."

Whose reasoning is correct according to Professor Lucas?

- **A.** Only Helen
- **B.** Only Kathy
- **C.** Both of them
- **D.** No one

Question: 39 of 40

Observe the following square braces containing four numbers.

[4, 6, 8, 32]

[3, 5, 7, 22]

[2, 4, 6, 14]

[5, 7, 9, ?]

If all four numbers in each brace follow the same rule, which number can replace the question mark?

- **A.** 11
- **B.** 44
- **C.** 108
- **D.** 54

Question: 40 of 40

If Adam got sick, he won't be able to submit the assignment before the deadline.

If Adam does not submit the assignment before the deadline, he cannot get an A+.

Adam didn't get an A+.

That means ..

- **A.** Adam got sick.
- **B.** Adam didn't get sick.
- **C.** Adam didn't get sick and he didn't study well.
- **D.** No conclusion.

Preptive Prepare Better

Series C Reading Comprehension Trial Test

Name: ______________________________

Date: ______________________________

Direction for Questions: 1 to 5

Read the extracts below and answer the questions.

Extract A: The Advantages of Sleep Quality: Improving General Wellness

Sleep is an indispensable component of human existence, playing a fundamental role in preserving body and mind at their peak. The significance of quality sleep extends beyond mere duration, encompassing restorative qualities of the sleep. This extract aims to examine the manifold advantages that result from attaining regular restful slumber.

Cognitive function is positively impacted by quality sleep, which is one of its most notable benefits. In terms of memory consolidation, problem-solving capabilities, creativity, and general mental acuity, a mind that is adequately rested exhibits superior performance. Superior sleep quality improves cognitive capabilities by permitting the brain to engage in essential processes, including information processing and memory formation. There is a strong correlation between enhanced physical health and quality sleep. The body engages in critical processes of restoration and rejuvenation while sleeping. Sufficient sleep enhances the functionality of the immune system, thereby mitigating the likelihood of contracting diseases and facilitating a more expedited recuperation process. Additionally, adequate sleep of high quality plays a role in weight maintenance by regulating hormones that control appetite and satiety, thereby decreasing the probability of excessive eating or cravings.

Sleep's significance in relation to emotional health cannot be emphasized enough. Emotional stability and resilience are enhanced by sufficient sleep, which enables us to modulate our emotions, combat stress, and experience an overall improvement in our mood. On the contrary, protracted sleep deprivation has been linked to the development of anxiety, mood disorders, and heightened vulnerability to negative emotional states.

Obtaining slumber of satisfactory quality has the potential to greatly augment productivity and performance across multiple domains. A well-rested individual is more attentive, focused, and capable of delivering optimal performance in various domains, including academics, exercises, and work. Adequate sleep engenders the necessary vitality, cognitive clarity, and drive to confront mundane responsibilities with efficacy and productivity.

The adage that "beauty sleep" is not a myth is partially accurate. A sufficient amount of high-quality sleep is essential for promoting a more radiant and youthful appearance, reducing the appearance of dark circles and puffiness, and sustaining a healthy complexion. By facilitating the restoration of damaged cells, regeneration of skin tissue, and regulation of hormonal levels, sleep enhances the body's physical well-being and vitality.

Extract B : Exploring the Consequences and Dangers of Maintaining Alertness for Two Days

Sleep is an indispensable physiological requirement that is fundamental to the body's optimal operation and general welfare. Nevertheless, there are circumstances in which people may discover themselves awake for prolonged durations, such as when they complete arduous tasks or perform an all-nighter. This extract will examine the potential consequences and hazards associated with a two-day period of continuous wakefulness.

Two days of wakefulness can have a profound negative impact on cognitive function. Insufficiency of sleep impairs memory, concentration, attention, and decision-making. Cognitive performance deteriorates with the accumulation of sleep deprivation hours, resulting in challenges with information processing, sluggish reaction times, and compromised problem-solving abilities.

Emotions may be heavily influenced by prolonged periods of wakefulness. Elevated emotional reactivity, irritability, and mood fluctuations are frequent symptoms of sleep deprivation. People may potentially manifest symptoms such as agitation, anxiety, and depressive episodes. Emotional disturbances have the potential to exert adverse effects on interpersonal connections and general welfare.

Coordination and physical performance are impaired by sleep deprivation. Extended periods of wakefulness can lead to adverse effects such as diminished motor skills, compromised strength, and slowed reflexes. Participating in physically strenuous tasks or operating machinery without sufficient rest can present significant hazards to one's safety.

The immune system is weakened by sleep deprivation, rendering the body more vulnerable to infections and diseases. A lack of sleep for two consecutive days can have a substantial detrimental effect on the immune system, rendering individuals more susceptible to contracting infections such as the flu and cold. This can potentially result in enduring health consequences and impede the body's capacity to efficiently combat diseases.

Prolonged wakefulness is associated with a variety of health hazards. It has the potential to induce hypertension, tachycardia, and hormonal imbalance. There exists a correlation between chronic sleep deprivation and an elevated susceptibility to cardiovascular diseases, obesity, diabetes, and mental health disorders.

Although in certain circumstances, remaining conscious for two days may appear to be an endurance test or a sacrifice that is essential, it is critical to be aware of the potential dangers and repercussions. Extended sleep deprivation has detrimental effects on various aspects of health, including cognitive performance, emotional state, and physical well-being. Sleep should be regarded as a fundamental component of a healthy way of life, with the goal of establishing and adhering to a regular sleep routine that provides the body and mind with an opportunity to recharge and restore. Keep in mind that an individual who has adequate rest is more capable of managing obstacles, delivering optimal performance, and preserving overall health.

Question: 1 of 30

According to Extract A, what is one benefit of good quality sleep ?

- **A.** Improved physical appearance
- **B.** Increased stress levels
- **C.** Reduced cognitive function
- **D.** Impaired productivity

Question: 2 of 30

According to Extract A, what role does good quality sleep play in emotional well-being?

- **A.** It increases susceptibility to negative emotions
- **B.** It improves emotional resilience
- **C.** It exacerbates mood disorders
- **D.** It causes chronic stress

Question: 3 of 30

According to Extract A, what is one aspect of physical health that is influenced by good quality sleep?

- **A.** Increased risk of illnesses
- **B.** Unhealthy weight gain
- **C.** Weakened immune system
- **D.** Faster recovery from illnesses

Question: 4 of 30

According to Extract B, what are some physical effects of staying awake for 2 days?

- **A.** Enhanced motor skills and coordination
- **B.** Improved physical performance and strength
- **C.** Slower reflexes and reduced physical coordination
- **D.** Increased energy levels and physical endurance

Question: 5 of 30

According to Extract B, how does sleep deprivation impact the immune system?

- **A.** Boosts the immune response and strengthens immunity
- **B.** Decreases the risk of infections and illnesses
- **C.** Weakens the immune system and increases susceptibility to illnesses
- **D.** Does not affect the immune system

Direction for Questions: 6 to 8

Read the text below and answer the questions.

Implementing Strategies to Prevent Global Warming

Global warming is an imminent ecological emergency that presents substantial perils to the Earth and its entire biosphere. Increasing temperatures have far-reaching implications, such as the occurrence of extreme weather phenomena, the draining of glaciers, and the disappearance of species. Although a comprehensive approach at the governmental and industrial levels is necessary to resolve this complex issue, individual efforts are also of the utmost importance. This article will examine doable measures that every individual can implement to aid the worldwide endeavor of mitigating climate change and safeguarding the planet against its detrimental effects.

It is essential to transition to sustainable energy sources in order to decrease greenhouse gas emissions. Consider installing renewable energy sources, such as solar or wind power, in your residence. When these alternatives are unattainable, utility providers may offer renewable energy programs as an alternative. Endorsing the advancement and deployment of renewable energy sources contributes to the mitigation of global warming, fosters innovation, and generates environmentally sustainable employment prospects.

An important contributor to worldwide greenhouse gas emissions is the combustion of fossil fuels for transportation purposes. We can mitigate climate change and decrease our carbon footprint through deliberate decision-making. Whenever practicable, carpooling, public transportation, walking, or cycling should be considered. Furthermore, to optimize fuel economy, select energy-efficient vehicles and ensure they are correctly maintained. Not only does adopting these practices reduce emissions, but it also fosters a more health-conscious way of life.

In addition to reducing carbon emissions, energy conservation also helps save money. Energy consumption can be drastically reduced through the use of energy-efficient appliances, appropriate insulation, and turning off lights when not in use. Additionally, natural ventilation and thermostat adjustments can aid in energy conservation. Bear in mind that incremental progress at the individual level can yield substantial outcomes on a worldwide level.

Consumer decisions exert a significant influence on the environment. Choose products that are favorable to the environment, such as those manufactured using sustainable practices or consisting of recycled materials. When feasible, reduce, reuse, and recycle materials. Adopt a sustainable stance by selecting biodegradable alternatives to single-use plastic items and minimizing one's consumption of such items. Through practicing conscious consumption, we can contribute to the conservation of resources and the mitigation of pollution.

Carbon dioxide absorption by trees is a critical factor in the mitigation of global warming. Contribute to afforestation and reforestation initiatives through tree-planting campaigns and support for organizations committed to forest restoration. Promote responsible forestry practices and contemplate the use of timber products that are recycled or derived from sustainable sources.

Disseminate your understanding of global warming among acquaintances, relatives, and coworkers. Initiate dialogues, coordinate public consciousness initiatives, and exploit social media channels as means to increase awareness regarding the critical nature of the matter. Promoting sustainable practices and inspiring collective action are largely dependent on education.

A global endeavor is required to combat global warming, encompassing the participation of governments, communities, businesses, and individuals. By adopting the recommendations delineated in this article, every individual can contribute to the effort of preserving the planet's future and alleviating the impacts of climate change. We can create a sustainable world and protect it from the dangers of global warming by working together. It is imperative that we take immediate action in order to safeguard future generations.

Question: 6 of 30

According to the paragraph, how can individuals contribute to energy conservation in their daily lives?

- **A.** Supporting green energy programs offered by utility providers.
- **B.** Recycling and reusing products whenever possible.
- **C.** Engaging in tree-planting initiatives.
- **D.** Turning off lights when not in use and using energy-efficient appliances.

Question: 7 of 30

Why is spreading awareness and educating others about global warming important?

- **A.** It encourages others to engage in tree-planting initiatives.
- **B.** It helps to increase individual energy consumption.
- **C.** It inspires collective action and promotes sustainable practices.
- **D.** It supports the exploration of fossil fuel sources.

Question: 8 of 30

How can individuals practice sustainable consumption to address global warming?

- **A.** Choosing petrol powered vehicles and maintaining them properly.
- **B.** Participating in recycling initiatives and minimizing single-use plastics.
- **C.** Supporting organizations dedicated to restoring forests.
- **D.** Opting for green energy programs offered by utility providers.

Direction for Questions: 9 to 12

Read the poem below then answer the questions.

Spring and Fall
***by* Gerard Manley Hopkins**

Margaret, are you gríeving
Over Goldengrove unleaving?
Leaves like the things of man, you
With your fresh thoughts care for, can you?
Ah! as the heart grows older
It will come to such sights colder
By and by, nor spare a sigh
Though worlds of wanwood leafmeal lie;
And yet you wíll weep and know why.
Now no matter, child, the name:
Sorrow's spríngs are the same.
Nor mouth had, no nor mind, expressed
What heart heard of, ghost guessed:
It ís the blight man was born for,
It is Margaret you mourn for.

Question: 9 of 30

What does the line "It ís the blight man was born for" suggest about the nature of sorrow?

- A. Sorrow is natural and inevitable.
- B. Sorrow can be avoided with effort.
- C. Sorrow is temporary and fleeting.
- D. Sorrow is caused by external factors.

Question: 10 of 30

What is the significance of the line "Sorrow's spríngs are the same"?

- A. It emphasizes the beauty of spring.
- B. It suggests that sorrow is eternal and unchanging.
- C. It describes the changing seasons.
- D. It highlights Margaret's love for nature.

Question: 11 of 30

What is the speaker's message to Margaret in the poem's closing lines?

- A. The speaker urges Margaret to embrace her grief.
- B. The speaker advises Margaret to forget her sorrows.
- C. The speaker reassures Margaret that her grief is temporary.
- D. None of the above.

Question: 12 of 30

What does the poem imply about the relationship between age and the experience of sorrow?

- A. Sorrow becomes less significant with age.
- B. Sorrow deepens and becomes more understood with age.
- C. Sorrow has no meaning to children.
- D. Sorrow is incomprehensible to the elderly.

Direction for Questions: 13 to 19

Read the poem below then answer the questions.

Stanzas
***by* Emily Bronte**

Often rebuked, yet always back returning
To those first feelings that were born with me,
And leaving busy chase of wealth and learning
For idle dreams of things that cannot be:

To-day, I will seek not the shadowy region;
Its unsustaining vastness waxes drear;
And visions rising, legion after legion,
Bring the unreal world too strangely near.

I'll walk, but not in old heroic traces,
And not in paths of high morality,
And not among the half-distinguished faces,
The clouded forms of long-past history.

I'll walk where my own nature would be leading:

It vexes me to choose another guide:
Where the gray flocks in ferny glens are feeding;
Where the wild wind blows on the mountain side.

What have those lonely mountains worth revealing?
More glory and more grief than I can tell:
The earth that wakes one human heart to feeling
Can centre both the worlds of Heaven and Hell.

Question: 13 of 30

What is the overall tone of the poem "Stanzas" by Emily Bronte?

- **A.** Joyful and celebratory
- **B.** Melancholic and introspective
- **C.** Energetic and enthusiastic
- **D.** Suspenseful and mysterious

Question: 14 of 30

Which poetic device is used in the line "And visions rising, legion after legion"?

- **A.** Metaphor
- **B.** Alliteration
- **C.** Simile
- **D.** Personification

Question: 15 of 30

In the poem, what do the "gray flocks in ferny glens" symbolize?

- **A.** Wealth and prosperity
- **B.** Nature and simplicity
- **C.** Loneliness and isolation
- **D.** Historical figures and events

Question: 16 of 30

What does the speaker mean by "Its unsustaining vastness waxes drear"?

- **A.** The vastness of the world is inspiring and uplifting.
- **B.** The vastness of the world is overwhelming and depressing.
- **C.** The vastness of the world is filled with unreal dreams.
- **D.** The vastness of the world is joyful and lively.

Question: 17 of 30

What is the deeper meaning behind the line "More glory and more grief than I can tell"?

- **A.** The speaker is indifferent to both glory and grief.
- **B.** Life contains both moments of great joy and deep sorrow.
- **C.** The speaker is unable to experience either glory or grief.
- **D.** Glory and grief are only illusions in life.

Question: 18 of 30

What does the speaker mean by "I'll walk where my own nature would be leading"?

- **A.** The speaker will follow their own instincts and desires.
- **B.** The speaker will follow a predefined path chosen by others.
- **C.** The speaker will walk in a place devoid of nature.
- **D.** The speaker will walk only in heroic and historic places.

Question: 19 of 30

What does the speaker mean when they say, "The earth that wakes one human heart to feeling"?

- **A.** The earth has the power to make people feel emotions.
- **B.** The earth has no effect on human emotions.
- **C.** The earth can only be understood through logic and reason.
- **D.** The earth is a source of great wisdom.

Direction for Questions: 20 to 23

Read the extracts and then answer the questions.

Extract A:
The Egypt Pyramids are ancient marvels that have captivated the world for centuries. Built as tombs for pharaohs, these magnificent structures stand as a testament to the incredible engineering skills of the ancient Egyptians. Their precise construction and enduring beauty continue to intrigue archaeologists and tourists alike. Each pyramid tells a story of the great civilizations that once thrived along the Nile River. Today, they serve as a reminder of the rich history and cultural heritage of Egypt.

Extract B:
The Great Wall of China is an architectural masterpiece and one of the most iconic landmarks in the world. Stretching over thousands of miles, it stands as a symbol of the strength and resilience of the Chinese civilization. Built to protect the empire from invasions, it is a testament to the determination and ingenuity of the ancient Chinese people. Today, it serves as a popular tourist attraction, allowing visitors to marvel at its grandeur and explore the history that lies within its walls.

Extract C:
The London Eye is a modern marvel that has become an iconic symbol of the city's skyline. Standing tall on the banks of the River Thames, it offers breathtaking panoramic views of London. As one of the world's largest Ferris wheels, it attracts millions of visitors each year. Whether it's during the day or at night, a ride on the London Eye provides a unique perspective of the city's landmarks, including the Houses of Parliament, Buckingham Palace, and St. Paul's Cathedral. It has become an essential part of any London itinerary.

Extract D:
The Opera House is a cultural gem that showcases the beauty of performing arts. As an architectural masterpiece, it stands as a symbol of elegance and creativity. With its unique design and acoustics, it provides a perfect setting for captivating performances, be it opera, ballet, or theater. The Opera House serves as a hub for artists and performers from around the world, attracting audiences who appreciate the power of artistic expression. It is a place where emotions come alive, leaving lasting impressions on those fortunate enough to witness its grandeur.

Question: 20 of 30

Which of the extracts refers to a place for performing arts?

- **A.** Extract A
- **B.** Extract B
- **C.** Extract C
- **D.** Extract D

Question: 21 of 30

Which of the extracts refers to the tomb of pharaohs?

- **A.** Extract A
- **B.** Extract B
- **C.** Extract C
- **D.** Extract D

Question: 22 of 30

Which of the extracts mentions the place that provides a unique perspective of the city's landmarks?

- **A.** Extract A
- **B.** Extract B
- **C.** Extract C
- **D.** Extract D

Question: 23 of 30

Which of the extracts depicts the structure designed to protect from invasions?

- **A.** Extract A
- **B.** Extract B
- **C.** Extract C
- **D.** Extract D

Direction for Questions: 24 to 30

Seven sentences have been removed from the text. Choose from the sentences (A – H) the one which fits each gap (1 – 7). There is one extra sentence which you do not need to use.

The importance of a digital library

1.... They provide users with easy access to a vast collection of resources. This essay explores the benefits of using a digital library, including convenience, cost-effectiveness, and the ability to access information anytime, anywhere.
One of the primary advantages of digital libraries is the convenience they offer. **2....** With just a few clicks, users can search for and retrieve information on various subjects, saving valuable time and effort. Additionally, digital libraries often provide advanced search options, allowing users to filter results based on specific criteria, making the research process more efficient.

3.... No longer limited by physical space, digital libraries enable users to access information from anywhere in the world, as long as they have an internet connection. This accessibility is particularly beneficial for individuals in remote areas or those with limited mobility. Furthermore, digital libraries can be accessed 24/7, eliminating the restrictions imposed by traditional library opening hours. **4...**

Digital libraries offer significant cost savings compared to traditional libraries. **5....** In contrast, digital libraries can provide access to countless resources without the need for physical storage space. This reduces costs associated with building maintenance, staff salaries, and inventory management. Moreover, digital libraries may offer free access to certain resources or charge a fraction of the cost typically associated with purchasing physical copies, making knowledge more affordable and accessible to a wider audience.

6..... Users can interact with peers, educators, or experts by sharing resources, discussing ideas, or engaging in virtual study groups. This collaborative environment fosters knowledge sharing and networking, providing opportunities to learn from diverse perspectives. Additionally, digital libraries facilitate the annotation and highlighting of texts, allowing users to personalize their reading experience. These features not only promote active learning but also enable users to easily refer back to specific sections of a text, enhancing comprehension and knowledge retention.

7... They are, convenience, accessibility, cost-effectiveness, and collaborative features. As technology continues to advance, digital libraries are becoming increasingly vital in providing users with a wealth of information and resources. Embracing this digital shift can revolutionize the way we access and engage with knowledge, ultimately benefiting individuals and society as a whole.

A. ***Unlike traditional libraries, which require physical visits, digital libraries allow users to access resources from the comfort of their own homes.***
B. ***Another crucial aspect of digital libraries is their accessibility.***
C. ***Physical books require substantial investments in purchasing, storing, and maintaining the collection.***
D. ***Digital libraries often offer collaborative features that enhance the learning experience.***
E. ***This flexibility allows users to study or conduct research at their own pace, fitting into their busy schedules.***
F. ***Embracing this digital shift can revolutionize the way we access and engage with knowledge, ultimately benefiting individuals and society as a whole.***
G. ***In the age of technology, traditional libraries are transforming into digital libraries.***
H. ***Digital libraries offer numerous advantages.***

Question: 24 of 30

Which one fits in (1) ?

- **A.** Sentence A
- **B.** Sentence B
- **C.** Sentence G
- **D.** Sentence H

Question: 25 of 30

Which one fits in (2) ?

- **A.** Sentence A
- **B.** Sentence B
- **C.** Sentence C
- **D.** Sentence D

Question: 26 of 30

Which one fits in (3) ?

- **A.** Sentence A
- **B.** Sentence D
- **C.** Sentence B
- **D.** Sentence E

Question: 27 of 30

Which one fits in (4) ?

- **A.** Sentence A
- **B.** Sentence B
- **C.** Sentence H
- **D.** Sentence E

Question: 28 of 30

Which one fits in (5) ?

- **A.** Sentence A
- **B.** Sentence B
- **C.** Sentence C
- **D.** Sentence D

Question: 29 of 30

Which one fits in (6) ?

- **A.** Sentence A
- **B.** Sentence B
- **C.** Sentence C
- **D.** Sentence D

Question: 30 of 30

Which one fits in (7) ?

- **A.** Sentence A
- **B.** Sentence G
- **C.** Sentence H
- **D.** Sentence C

Preptive Prepare Better

Series C Mathematical Reasoning Trial Test

Name: ______________________

Date: ______________________

Question: 1 of 35

Francisco needs to do quite a few errands.
He needs to stop by the dry cleaners, the post office, the hardware store, and the bank. There are a number of ways he can tackle his list of stops.

How many ways can he accomplish his tasks?

- **A.** 10
- **B.** 12
- **C.** 16
- **D.** 24

Question: 2 of 35

All the friends gathered at Teresa's home for a party. Mila came 5 minutes before the party starts. The party started at 20:00. Mila took 1 hr 5 minutes to drive from her home to Teresa's home.

When did she leave her home?

- **A.** 18:50
- **B.** 18:55
- **C.** 17:55
- **D.** 17:50

Question: 3 of 35

Donald, Tom and Lily had a total of 640 arrows.
Donald had 68 more arrows than Tom.
Lily had twice as many arrows as Donald.

How many arrows did Tom have?

- **A.** 108
- **B.** 109
- **C.** 210
- **D.** 354

Question: 4 of 35

A box measures 56 cm long, 23 cm wide and 106 cm high. Elise wants to fill the box with as many cubes as she can.
Each cube has a volume of 64 cu. cm.
What is the maximum number of cubes that Elise can fill the box with?

- **A.** 1110
- **B.** 1860
- **C.** 1820
- **D.** 2000

Question: 5 of 35

Jason finished reading all 185 pages of a story book in 5 days. For each day, he read 10 pages more than the previous day.
How many pages did he read on the first day?

- **A.** 17
- **B.** 18
- **C.** 23
- **D.** 35

Question: 6 of 35

Select the suitable letters for the empty boxes.

A	B	C	D
F	E	D	(i)
E	F	G	(ii)
J	I	H	G

- **A.** E, H
- **B.** F, H
- **C.** C, H
- **D.** C, F

Question: 7 of 35

You can see 11 pillars of its fence when you are looking from each of the three sides of an equilateral triangular land. How many pillars are there in the fence?

- **A.** 33
- **B.** 27
- **C.** 30
- **D.** 31

Question: 8 of 35

You need to make the following shape a square. How many more small squares will you need minimum ?

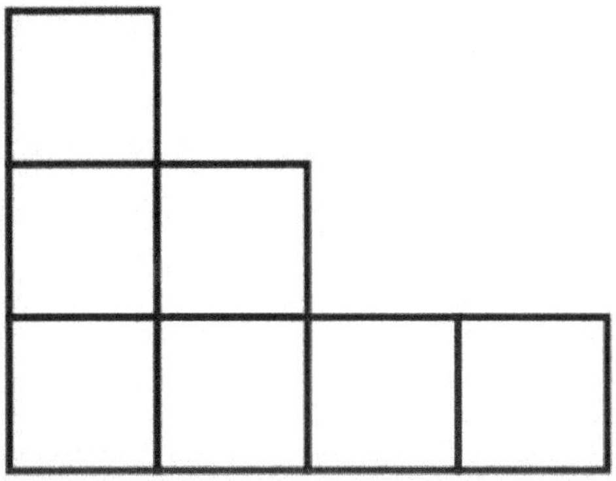

- A. 6
- B. 7
- C. 9
- D. 10

Question: 9 of 35

Jennie is in a bus queue. There are 4 people behind her and 6 people in front of her. If Jennie's brother was in the middle of the queue, which of the following is correct?

- A. He is the 5th of the queue
- B. He is the 6th of the queue
- C. He is the 7th of the queue
- D. He is the 8th of the queue

Question: 10 of 35

David has to go 1400 m to school. Yesterday he left home for school and after riding his bicycle for 600 m, he remembered that he forgot the water bottle. So, he returned home and again left for school. How far has David gone?

- A. 2000 m
- B. 2600 m
- C. 3200 m
- D. 1400 m

Question: 11 of 35

Siti wanted to buy 30 markers but she was short of $14. She bought 22 markers instead and had $18 left. How much money did Siti have?

- A. $4
- B. $32
- C. $106
- D. $18

Question: 12 of 35

Kavya bought 28 mangoes and pomelos. Each mango cost $3 and each pomelo cost $5. She paid a total of $118 for all the fruits.
How many pomelos did she buy?

- **A.** 17
- **B.** 13
- **C.** 32
- **D.** 89

Question: 13 of 35

The figure below is a net.

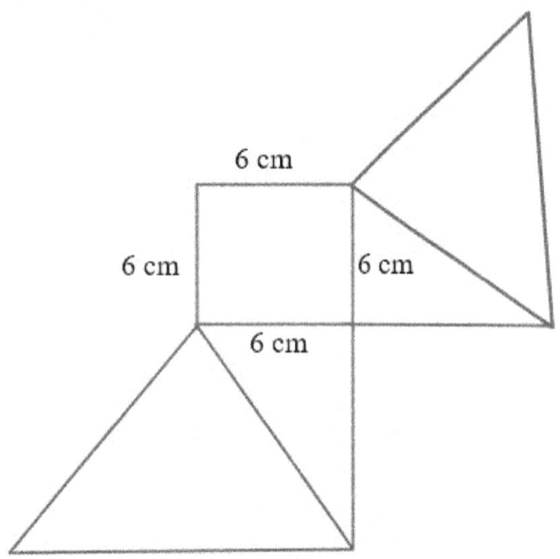

If the net is folded, what object is going to be formed?

- **A.** Cuboid
- **B.** Cylinder
- **C.** Pyramid
- **D.** Cone

Question: 14 of 35

Four points P, Q, R, and S are such that PS is parallel to QR and PQ is parallel to RS. Show the possible positions of P, Q, R, and S.

A.

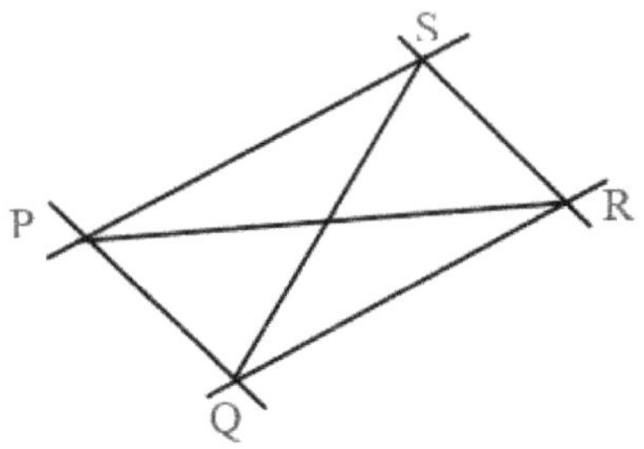

B.

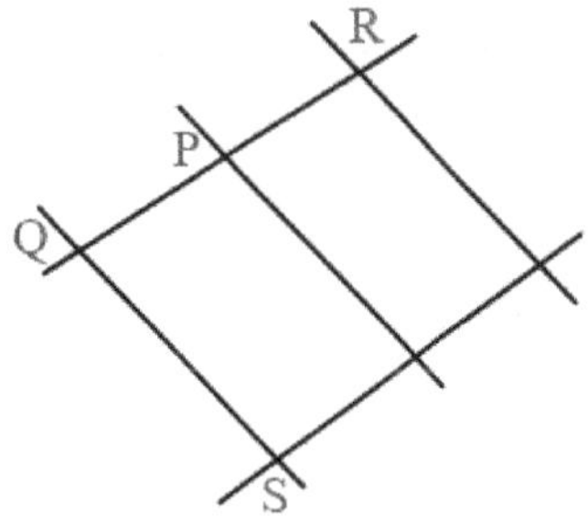

C.

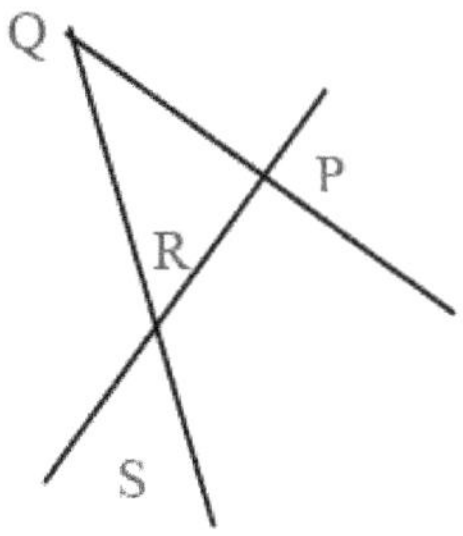

D.

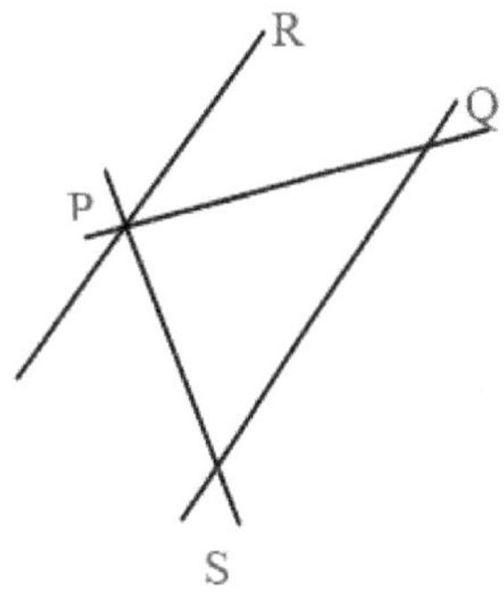

Question: 15 of 35

Point M $\left(\frac{1}{2}, 1\right)$ is the mid-point of points A (a, -3) and B (4, b). Find a and b.

A. -2, 1

B. -3, 5

C. 3, -1

D. 5, 3

Question: 16 of 35

The table below shows the average ages that people live in different countries. This is called 'life expectancy'. Work out the range for men and the median life expectancy for women.

Country	**Life expectancy**	
	Men	Women
UK	72	78
Britain	43	46
Australia	62	68
Germany	57	61
Poland	75	80
France	63	66

A. Men range = 57, Women median = 46

B. Men range = 43, Women median = 46

C. Men range = 32, Women median = 67

D. Men range = 72, Women median = 68

Question: 17 of 35

Jenny entered a diving competition. Each of the dives that Jenny makes has a dive rating.
The harder the dive the higher the rating. Here are the rules for working out a score:

i. Remove the highest and the lowest mark
ii. Add together the middle three marks
iii. Multiply the answer of (ii) by the dive rating.

Each mark is out of 10. Jenny dives with a rating of 3.25. The judge gives the marks: **7.5, 7.7, 7.9, 7.3, 7.6**

Work out Jenny's score.

- **A.** 80.2
- **B.** 76.7
- **C.** 74.1
- **D.** 53.4

Question: 18 of 35

The diagram below represents two flashing lamps A and B. The lamps are switched on at the same time. Work out how often the two lamps flash together.

Lamp A flashes 20 seconds.

Lamp B flashes 35 seconds

- **A.** 100 seconds
- **B.** 140 seconds
- **C.** 200 seconds
- **D.** 160 seconds

Question: 19 of 35

Jason and Mark each think of a prime number that is greater than 100 and less than 110.
The total **sum of the digits** of Jason's number is 4 and the total sum of the digits of Mark's number is 10.

What number do you get when Jason's number and Mark's number are added together?

- **A.** 203
- **B.** 212
- **C.** 214
- **D.** 109

Question: 20 of 35

Ravi bought 18 more egg tarts than Ayesha. Jessie bought 3 times as many egg tarts as Ravi. All the three girls bought a total of 187 egg tarts.
How many egg tarts did Ravi and Ayesha buy altogether?

- **A.** 55
- **B.** 64
- **C.** 60
- **D.** 70

Question: 21 of 35

An eraser cost 30 cents and a packet of 5 erasers cost $1. Jonathan bought exactly 27 erasers.
What was the least amount of money that Jonathan spent on the erasers?

- **A.** $5.50
- **B.** $4.00
- **C.** $5.60
- **D.** $6.00

Question: 22 of 35

The mass of 2 identical packets of salt and 3 identical packets of sugar is 3862 g. The mass of 15 such packets of salt and 3 such packets of sugar is 8490 g.
What is the mass of one such packet of salt (in gms)?

- **A.** 356
- **B.** 348
- **C.** 376
- **D.** 329

Question: 23 of 35

The table below shows the rate of parking a car in a car park.

Parking Charges	
First Hour	$2.50
Every additional half hour or part thereof	$1.20

How much does Mr Bhim need to pay for parking his car from 7.30 a.m. to 9.15 a.m.?

- **A.** $7.80
- **B.** $6.50
- **C.** $5.00
- **D.** $4.90

Question: 24 of 35

A company ordered 25 potted plants to line up along their corridor from one end to the other end at equal spacing of 1.2 m apart. 4 of the potted plants broke during delivery and could not be used. The company used the remaining potted plants to line up along their corridor from one end to the other end at a new equal spacing. What was the new equal spacing between two potted plants?

- A. 2 m
- B. 1.44 m
- C. 1.64 m
- D. 1.50 m

Question: 25 of 35

The average height of a group of girls was 155 cm.
When 15 boys joined the group, the average height became 158 cm.
Given that the average height of the 15 boys was 165 cm, find the number of girls in the group.

- A. 45
- B. 25
- C. 35
- D. 75

Question: 26 of 35

The following pie graph shows the number of certain players of a particular country playing different games. Study the pie graph carefully and answer the following questions:

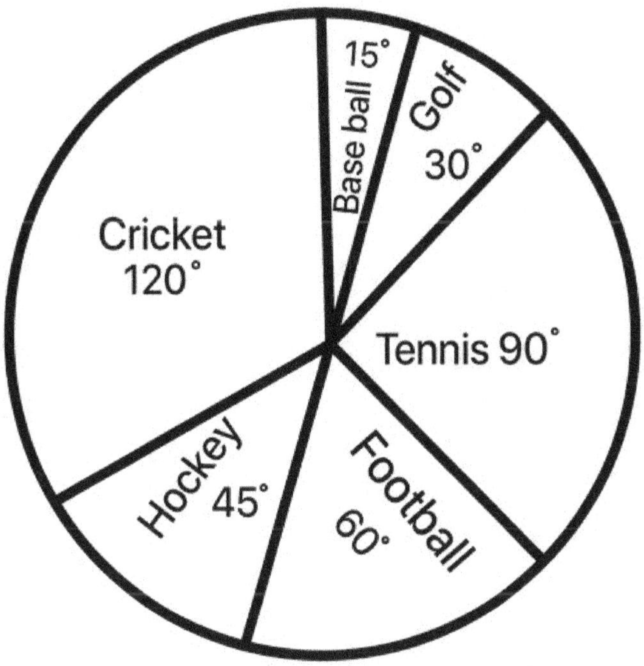

If the total number of persons is 90,000, then the difference between the number of persons playing only Tennis and Hockey is ...

- **A.** 11,250
- **B.** 8,250
- **C.** 7,500
- **D.** 6,750

Question: 27 of 35

A shipping company charges customers based on the weight of their packages. The weight **categories** and corresponding rates are as follows:

Upto 5 kg: $10
5.1 kg to 10 kg: $15
10.1 kg to 15 kg: $20
15.1 kg and above: $25

If a customer's package weighs 12.5 kg, how much will they be charged?

- **A.** $10
- **B.** $15
- **C.** $20
- **D.** $25

Question: 28 of 35

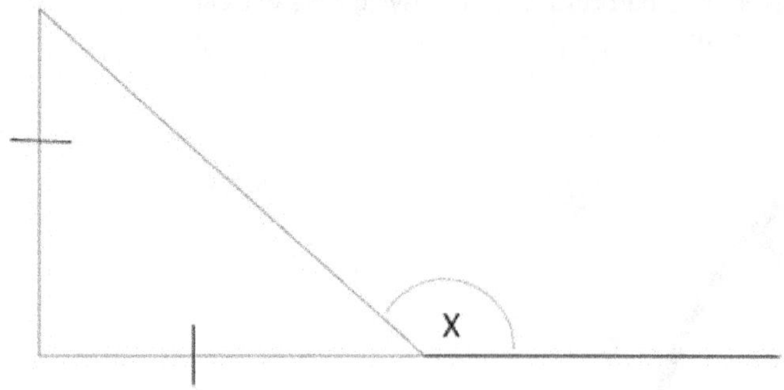

Given above is a right angled isosceles triangle. What is the value of x?

- A. 135 °
- B. 125 °
- C. 120 °
- D. 115 °

Question: 29 of 35

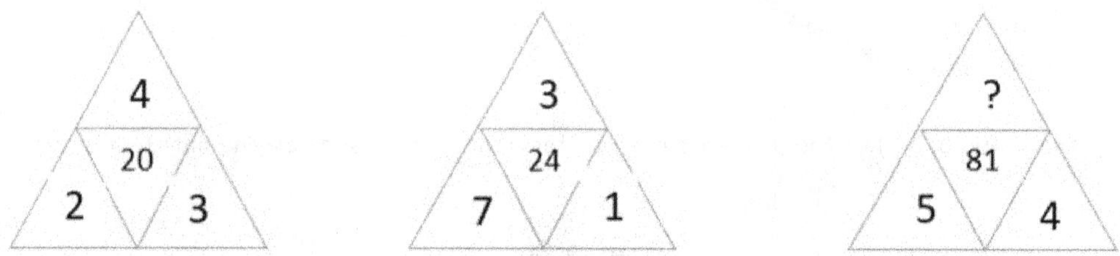

Identify the pattern and replace '?' mark.

- A. 2
- B. 6
- C. 9
- D. 13

Question: 30 of 35

Handmade paper flowers are put in a bag. They are all identical in shape. There are 6 yellow flowers, 9 white flowers, 12 red flowers and 8 blue flowers.
What is the probability of taking out a flower, which is yellow, blue, white or red?

- A. 0
- B. $^{4}/_{35}$
- C. $^{1}/_{35}$
- D. 1

Question: 31 of 35

Below pie chart illustrates favorite sports of grade 7 students. Total no. of students is 150.

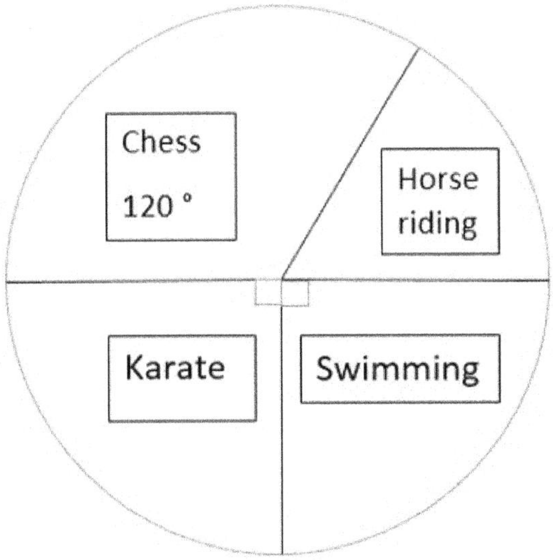

How many students like chess?

- A. 50
- B. 60
- C. 70
- D. 80

Question: 32 of 35

In a house there is a huge window which has a height of 3m and width of 4m. Amelia decides to paste a painting on it leaving a border of 50cm wide all around it. What is the area of the border?

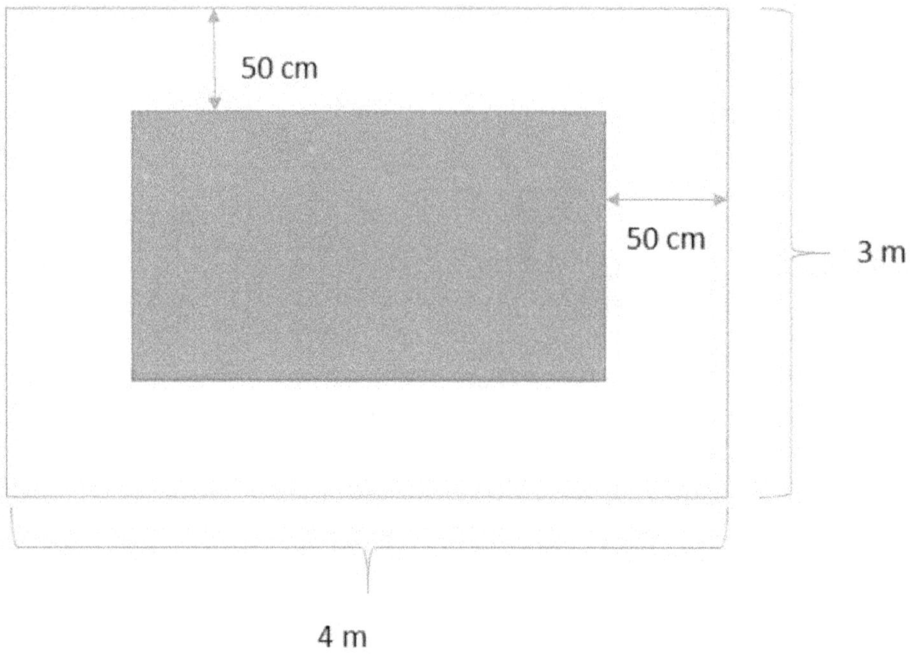

A. $12\ m^2$

B. $8\ m^2$

C. $6\ m^2$

D. $4\ m^2$

Question: 33 of 35

Joe and Alan had a total of 33 marbles at first. After Joe lost 3 of his marbles and Alan lost half of his marbles, they had an equal number of marbles left.
How many marbles did Alan have in the end?

A. 14

B. 10

C. 30

D. 12

Question: 34 of 35

Carol, Dora, and Eve planted 11,753 roses altogether. Carol planted half as many roses as Dora and Eve planted half as many roses as Carol.
How many more roses did Dora plant than Eve?

A. 5073

B. 5037

C. 5057

D. 5017

Question: 35 of 35

Kelly took 4 hours to paint a room. Richard took 8 hours to paint an identical room. If both of them worked together, how long (**in hours**) would they take to paint the room?

- **A.** $2\ ^{2}/_{3}$ hrs
- **B.** $2\ ^{1}/_{3}$ hrs
- **C.** $4\ ^{1}/_{3}$ hrs
- **D.** 2 hrs

Preptive Prepare Better

Series C Thinking Skill Trial Test

Name: ______________________________

Date: ______________________________

Question: 1 of 40

If Mary goes to the beach she will play with sand.
If she plays with sand she might collect sand into a bucket to take home.
If she has sea sand she will design wall art under the theme 'beach'.
Only if she designs satisfying wall art she will exhibit it in the art gallery.

If the given information is true, which one of the following is not correct?

- **A.** If Mary designs wall art under the theme 'beach' her art must be exhibited in the art gallery.
- **B.** If Mary goes to the beach she might design a wall art under the theme beach.
- **C.** If Mary's art was exhibited in the art gallery then she must have designed a satisfying wall art.
- **D.** Even though Mary didn't go to the beach her art was exhibited in the art gallery.

Question: 2 of 40

Both Maya and Rima are gifted students. Maya is especially good at Math. Both of them are very good at science. Both of them got selected for the final rounds of Math and science quiz competitions representing their school, along with other students.

Students from their school won both Math and science best prizes.

If the above information is true, which of the following is correct?

- **A.** If Maya is not the student who won the Math prize, definitely Rima is not the person who won that prize.
- **B.** Maya must have won the Math prize and Rima have won the science prize.
- **C.** Both Maya and Rima have won prizes.
- **D.** Neither Maya nor Rima might win any prize in quiz competitions.

Question: 3 of 40

The vet clinic opens at morning 8 O' clock every weekend.
Amalia and Robinson came to the clinic with their pets, but it was closed.

Amalia: "As today is Sunday and it's 8:20 a.m. the clinic should be open by now."

Robinson: "Today the veterinary surgeon must be on a holiday, let's go back."

If the information is true, whose expecation is **most likely** be correct?

- **A.** Amalia only
- **B.** Robinson only
- **C.** Both Amalia and Robinson
- **D.** Neither Amalia nor Robinson

Question: 4 of 40

Every mathematician has a higher IQ.
Neil is a mathematician.

So, we can conclude that ...

- A. Neil has a higher IQ.
- B. Neil's IQ can not be measured.
- C. Neil doesn't have a higher IQ.
- D. No conclusion

Question: 5 of 40

Below are four statements. There may be additional statements that may not be required to arrive at a conclusion.

"Shipowners are rich."
"Rich people have SUVs."
"Kevin has an SUV."
"New SUV models are mostly hybrid."

Which is the most correct conclusion?

- A. Kevin is surely a shipowner
- B. Kevin is surely not a shipowner
- C. Kevin might be a shipowner
- D. Kevin is rich but not a shipowner

Question: 6 of 40

An astronaut must have excellent observation skills and the ability to work under intense pressure.

If the given information is correct which of the following must be true?

- A. If Andrew has both excellent observation skills and the ability to work under intense pressure he must be an astronaut
- B. If Andrew has only the ability to work under intense pressure he cannot be an astronaut.
- C. If Andrew has only the ability to work under intense pressure he might be an astronaut
- D. If Andrew has excellent observation skills he must be an astronaut

Question: 7 of 40

According to new studies, all octopuses and cuttlefish are venomous. But **Blue-ringed octopuses remain the only one dangerous to humans**. Blue-ringed octopuses can kill humans by biting and injecting venom.

If the information given is true, which of the following must be true?

- **A.** If an animal died because of an octopus attack, that octopus must be a Blue-ringed octopus
- **B.** None of the humans can be killed by an octopus attack
- **C.** If a human died as a result of an octopus attack that octopus must be a Blue-ringed octopus
- **D.** According to new studies, people should stop eating cuttlefish as they are venomous.

Question: 8 of 40

A description of the four students who scored the highest marks in each class is given below.

- *Sam scored twenty marks more than Kai*
- *Roy scored twenty-one marks less than Sam*
- *Henry scored two marks more than Roy*

Based on the above information, which one of the following must be true?

- **A.** Henry is the second.
- **B.** Roy is the first.
- **C.** Sam is the third.
- **D.** Neither Sam nor Kai is the first.

Question: 9 of 40

In Monica's company :

Anyone who has a printer at their home has a laptop.
Anyone who has an electric car has a solar panel at their home.
Sam and Robin are also working in Monica's company.

Jack: "Sam has a solar panel in his house but he doesn't have an electric car."

Patricia: "Robin **used to have a printer** but he does not have a laptop"

If the given information is correct, whose reasoning is **impossible**?

- **A.** Jack only
- **B.** Patricia only
- **C.** Both Jack and Patricia
- **D.** Neither Jack nor Patricia.

Question: 10 of 40

Identify reasonable conclusion(s) following the statement.

Statement: For all the TV sets manufactured in Australia, the 'Solar' brand has the largest sale.

Conclusion I: Volume of sales of all brands of TV sets manufactured is known.
Conclusion II: The production of no other TV set is as large as that of 'Solar'.

- **A.** Only conclusion I follows
- **B.** Only conclusion II follows
- **C.** Neither conclusion I nor conclusion II follows
- **D.** Both the conclusion follows

Question: 11 of 40

Please refer to the statements below :

A: Kathy's cakes are tasty but expensive.
B: She gets fewer orders than other cake shops.

Which of the following shows the relation between A and B?

- **A.** A is the effect and B is the cause
- **B.** A is the cause and B is the effect
- **C.** A and B are effects of a different cause
- **D.** A and B are independent

Question: 12 of 40

Observe the following patterns of numbers.

- [2, 4, 5, 22]
- [6, 7, 8, 62]
- [8, 6, 5, A]

Based on the pattern, what is value of A ?

- **A.** 53
- **B.** 38
- **C.** 30
- **D.** 48

Question: 13 of 40

John had just finished his Selective test. He told his friends :
"Some questions in the exam were difficult."

Which of the following has the **opposite meaning** of what John had said ?

- **A.** Some questions in the exam were easy.
- **B.** All the questions in the exam were difficult.
- **C.** Some questions in the exam were not difficult.
- **D.** All the questions in the exam were easy.

Question: 14 of 40

Statement : "All engineering students should have a 6-month training experience before getting a job."

Neville and Smith are *providing their opinions* based on the above statement.

Neville: "Engineering students don't have practical skills before training."
Smith: "Training help students to gain some good knowledge and skills related to the industry."

Who is most correct according to statement ?

- **A.** Only Smith is correct
- **B.** Only Neville is correct
- **C.** Both are correct
- **D.** Both are wrong

Question: 15 of 40

Students in group A are good football players and athletes. Students in group B are not into athletics but they are good at cricket. Group C students are good at all the sports except athletics. There are no other groups.

Which of the following are correct about the students?

- **A.** Group A and C are only good at athletics
- **B.** Only two groups are not good at cricket
- **C.** Only one group has are good cricketers
- **D.** Only two groups have good football players

Question: 16 of 40

Karen is an electrical engineering student. During Lab time, he told is friend Rosy :
"If your circuit is not working, probably there is a short circuit, or the IC is blown."

But later Rosy found the error and Karen was proven wrong.
So, which is the most correct conclusion from the situation ?

- **A.** There were no short circuits
- **B.** The IC wasn't blown
- **C.** There were no short circuits, and the IC wasn't blown
- **D.** There were no short circuits, but IC can be either blown or not.

Question: 17 of 40

You are given a square shaped paper to make a hole in the middle as shown in the figure.

Minimally, how many times do you need to fold the paper to make the hole with a single cut?

- **A.** 1
- **B.** 2
- **C.** 3
- **D.** 4

Question: 18 of 40

Linda : "Every photographer has an expensive camera. I know that Mary doesn't have an expensive camera."

Which one of the below may be Linda's conclusion ?

- **A.** Mary is not a photographer
- **B.** If Mary buys an expensive camera, she will become a photographer
- **C.** No conclusion can be made
- **D.** None of the above

Question: 19 of 40

The following statements are true.

a. The square of X is 16
b. Y=0
c. L=2XY

When finding the value of L, which statement(s) is/are necessary?

- **A.** a only
- **B.** b only
- **C.** Neither a nor b
- **D.** Both a and b

Question: 20 of 40

A dice has faces labeled 1 to 6. Which statement is necessarily true if 2 is adjacent to 1, 3, and 5?

- **A.** 1 is opposite to 5
- **B.** 2 is adjacent to 4
- **C.** 4 is adjacent to 6
- **D.** 4 is opposite to 6

Question: 21 of 40

James walked northwards for 20 meters. Upon turning left, he went for 40 meters. James then turned left again, walking for 20 meters again. Lastly, he turned right and walked for 20 meters. How far had he traveled from where his journey began?

- **A.** 40
- **B.** 100
- **C.** 80
- **D.** 60

Question: 22 of 40

Among A, B, C, D and E, A is taller than B but shorter than C. B is taller than only E . C is not the tallest.

Who among them will be in the middle if they stand in the order of their heights?

- **A.** A
- **B.** C
- **C.** B
- **D.** E

Question: 23 of 40

Identify implicit assumption(s) behind the statement.

Statement: "Those who are appearing for this examination for the first time, should be helped in filling up the forms". - An instruction to the invigilating staff.

Assumption I: The form is somewhat complicated.
Assumption II: Candidates can appear more than once for this examination.

- **A.** Only Assumption I is implicit
- **B.** Only Assumption II is implicit
- **C.** Neither Assumption I nor II is implicit
- **D.** Both the assumptions are implicit.

Question: 24 of 40

If Sara enters her dream university, she will leave her country.
If Sara leaves her country, she will have to cook for herself on her own.

If the above information is true, which of the following cannot be true?

- **A.** Sara left her country, but she didn't enter her dream university
- **B.** Sara cooked for herself on her own but she didn't enter her dream university
- **C.** Sara entered her dream university but she didn't cook for herself on her own
- **D.** Sara cooked for herself on her own but she didn't leave her country.

Question: 25 of 40

Mr. Leon promised Jason that he will deposit money to Jason's account before 4 p.m. but only if he gets some free time.

Jason: "Now it's 10 p.m. Mr. Leon still hasn't deposited money, he must not have gotten free time"

Nicole: "If he deposited money before 4 p.m. then definitely he has got the free time."

If the information box is true, whose reasoning is not correct?

- **A.** Jason only
- **B.** Nicole only
- **C.** Both Jason and Nicole
- **D.** Neither Jason nor Nicole

Question: 26 of 40

Indian paradise flycatchers are beautiful bird whose male have elongated central tail feathers. Their heads are glossy black with a black crown and crest.

Adam: "If a male bird has elongated central tail feathers, that bird must be an Indian paradise flycatcher"

Hanna: "If a bird's head is glossy black with a black crown and crest, it must be an Indian paradise flycatcher"

If the information is true, whose reasoning is correct?

- **A.** Adam only
- **B.** Hanna only
- **C.** Both Adam and Hanna
- **D.** Neither Adam nor Hanna

Question: 27 of 40

Mr. Roger is an investigator. He knows about everyone, every criminal and notorious people in his area where he lives in. He is living there for 20 years. He has a special ability to memorize anyone's face or actions.

Mr.Roger: "Just show me the CCTV video of the thief. I will tell exactly who is he by his movements!"

Which one of the following sentences, if true, shows the mistake Mr. Roger has made?

- **A.** He can't memorize all the faces and movements of every person in the area for such a long time.
- **B.** He can't remember every movement of a single person.
- **C.** He does not know about the thieves who do not live or came before into his area.
- **D.** He can only give us predictions, he can't prove who is the thief.

Question: 28 of 40

Archaeopteryx is an iconic fossil, often thought of as the 'missing-link' between dinosaurs and birds. They are sometimes referred to by its German name, "Urvogel", an extinct genus of bird-like dinosaurs.

A group of friends is visiting a huge forest with a guide.

Guide: "Don't go here and there alone as there might be very dangerous animals like bears and urvogels"

Which of the following shows what is wrong with the guide's advice?

- **A.** If they are not allowed to walk freely, what is the use of coming to a forest, its better if they stay at home instead.
- **B.** Walking in that dangerous forest cannot guarantee safety.
- **C.** Archaeopteryx is an extinct animal so they can't be in that forest.
- **D.** There are many more dangerous things in the forest.

Question: 29 of 40

Nitrogen is an essential element for plant function and growth. Plants get the nitrogen that they need from the soil. Nepenthes is an insectivorous plant. Insectivorous plants have special leaves which can trap insects and kill them to derive nitrogen from the insects.

If the information is correct, which of the following statement is correct?

- **A.** Insectivorous plant might grow in nitrogen-deficient soil.
- **B.** Nitrogen is important for every plant, therefore Nepenthes can't grow in nitrogen-deficient soil.
- **C.** If a plant has special leaves which can trap insects and kill them that must be a Nepenthes plant.
- **D.** As nepenthes is an insectivorous plant it does not need nitrogen to grow.

Question: 30 of 40

Global warming and air pollution are serious threats to life on earth. Vehicle engines are the main reason for both air pollution and increasing global warming. If we reduce using vehicles by a significant amount, we can reduce air pollution too. For that, the government has to legislate laws and rules to completely stop private vehicle production, like cars, vans, jeeps, etc., and encourage people to use public transportation. Along with that, the government has to introduce a system to recycle private vehicles and manufacture bicycles, which are friendly for both environment and health. So people can travel short distances by bicycle and long distances by bus or train or any other public transport method.

Which of the following sentences most weakens the above argument?

- **A.** The air is polluted by public transport methods too.
- **B.** Recycling private vehicles and manufacturing bicycles is a waste of money.
- **C.** Private vehicles can be colourful.
- **D.** Cycling stimulates and improves your heart, lungs and circulation, reducing your risk of cardiovascular diseases.

Question: 31 of 40

The time remaining to finish 100 pages of the book is one week (7 days). Which of the following will ***surely*** make you miss your deadline?

- **A.** Reading 20 pages per day for five days
- **B.** Reading 25 pages every other day
- **C.** Reading 30 pages per day for the last three days
- **D.** Reading 10 pages each day for the whole week

Question: 32 of 40

Incandescent bulbs are having a tungsten coil. Also, their power consumption is high.
However, LED bulbs are totally the opposite.

Based on above information, LED bulbs ....

- **A.** Do not have a tungsten coil.
- **B.** Do not have a tungsten coil but the power consumption is high.
- **C.** The power consumption is low but they don't have a tungsten coil.
- **D.** Don't have a tungsten coil or the power consumption is low.

Question: 33 of 40

Carpenter: "If it doesn't rain I can finish your work by next Monday."

Robin: "Are you sure? It is very hard to believe."

What does Robin think?

- **A.** If it doesn't rain the Carpenter will finish the work by next Monday.
- **B.** It will rain and Carpenter will finish the work by next Monday.
- **C.** Carpenter will finish the work by this Monday.
- **D.** It won't rain and Carpenter will not finish the work by next Monday.

Question: 34 of 40

If there is heavy rainfall for two days, there is a risk of flood in Kevin's hometown.
If there is a risk of flood, people have to move to a safe place.

Kevin said it was heavily raining there since the day before yesterday. Which of the following is correct based on information above?

- **A.** Kevin's family has to move to a safer place.
- **B.** Kevin's family doesn't need to move to another place.
- **C.** No conclusion can be made.
- **D.** None of the above.

Question: 35 of 40

I said : "*If it's raining, we cannot play cricket. If the bat is broken, we cannot play cricket either.*"

Rickey said he bought a new bat today.

Which of the following is correct about the given scenario?

- **A.** We cannot play cricket.
- **B.** Rain is the only reason we won't play cricket.
- **C.** We can play cricket even if it rains.
- **D.** None of the above.

Question: 36 of 40

Statement : If it is spring, farmers start farming.

Based on the **above statement alone,** which ones of the below must be true ?

- **A.** If it is not the spring, farmers don't farm.
- **B.** If farmers are farming, it is spring.
- **C.** If farmers are not farming, it is the winter.
- **D.** If farmers are not farming, it may not be the spring.

Question: 37 of 40

All Italian restaurants have tasty food.
James had a meal in ABC restaurant and he said it was tasty.

What can be concluded about ABC resturant based on above two statements ?

- **A.** ABC is an Italian restaurant.
- **B.** ABC isn't an Italian restaurant.
- **C.** Cannot decide.
- **D.** ABC resturant is closed.

Question: 38 of 40

If the sun is so shiny and the rainfall is very low, the lakes will dry up.
The sun is so shiny these days.

Which of the following is the most correct conclusion regarding the above scenario?

A. If the rainfall is very high the lakes won't dry up.

B. The lakes will dry up.

C. The lakes won't dry up.

D. If the rainfall is very low the lakes will dry up.

Question: 39 of 40

If the meaning of "(Thomas)@ (restaurant) * (Kamala) = (dinner) $ (7 O' clock)" is "Thomas went to a restaurant with Kamala for dinner before 7 O' clock", what is the meaning of "(Michael) @ (cinema) $ (film starts)"?

A. Michael goes to the cinema before it starts.

B. Michael went to the cinema before the film starts.

C. Michael went to the cinema and watched a movie.

D. Michael went to the cinema with the film starts.

Question: 40 of 40

Sam : *If a country is economically stable, people have money.*

Ram : *If people have money, they celebrate many festivals.*

Sam and Ram are both correct. However, in some countries, people doesn't celebrate many festivals. That means...

A. Some countries are economically stable.

B. Some countries are economically unstable.

C. Cannot decide.

D. Economy is not an important aspect for a country.

Preptive Prepare Better

Series D Reading Comprehension Trial Test

Name: ______________________________

Date: ______________________________

Direction for Questions: 1 to 6

Read the extracts below and answer the questions.

Extract A : Understanding Cats: Our Feline Companions

Cats have enchanted people for generations with their charming grace and intriguing nature. These enthralling creatures have earned a particular place in our hearts and homes. Cats, being one of the most popular pets globally, bring us joy, companionship, and a touch of enchantment.

Cats are members of the Felidae family, a predatory lineage. Domestic cats' ancestors were solitary hunters, and they still have exceptional hunting instincts today. Cats are highly able to hunt a prey, with their acute vision, sensitive whiskers, nimble body and retractable claws. Though domesticated cats no longer rely on hunting for existence, their innate inclinations often manifest through playful behavior and occasional "gifts" given to their human friends.

Cats communicate by using vocalizations, body language, and scent marking. Cats rarely meow at each other, therefore meowing is mostly intended for communication between cats and people. Purring, on the other hand, can indicate contentment as well as stress. Observing their body language, such as tail posture, ear motions, and pupil dilation, might help us better comprehend their moods and needs.

Cats are known for their independence. Unlike dogs, they are more self-reliant and value their personal space and autonomy. However, this does not imply that cats are cold or unapproachable. Cats build strong ties with their human friends and frequently show affection through head bumps, kneading, and purring. Respecting their limits and providing a nurturing atmosphere aid in the development of trust and the strengthening of the human-cat relationship.

It is critical to provide an enriching environment for a cat's well-being. Cats flourish in environments that allow them to engage in play, exploration, and cerebral stimulation. Cat trees, scratching posts, interactive toys, and vertical areas allow them to engage in natural behaviors, keep healthy claws, and exercise. Additionally, designated resting and concealing locations provide cats with a sense of security and seclusion.

Meeting the physical and emotional needs of one's cat is part of being a responsible cat owner. This involves feeding a balanced food, getting regular veterinarian treatment, and making sure cat has access to fresh water and a clean litter box. Regular playtime and engaging sessions keep cats cognitively and physically occupied.

Cats have a long history of providing companionship, entertainment, and a sense of comfort to people. Understanding their natural inclinations, communication methods, and individual personalities enables us to form stronger links and offer a meaningful and enriching life for them. We have the privilege and responsibility of caring for these magnificent creatures as responsible cat owners. Let us appreciate the individuality of each feline companion and continue to honor the enduring bond between humans and cats.

Extract B: A hero cat saves the life of a heart attack victim, bolstering research that demonstrates pets may save people in distress

A brave cat in the United Kingdom saved his owner's life by slamming his paws on her chest to wake her up after she had a heart attack while sleeping. The astonishing behavior backs up previous study that reveals pets not only want to help their owners, but will even try to aid them on their own.

Sam Felstead, 42, was sleeping when her 7-year-old cat Billy woke her up at 4:30 a.m. at her Nottingham, England home. She discovered she couldn't move her body and was experiencing shooting pains down her right side, so she summoned her mother for assistance. Felstead, a receptionist at Nottingham's Queen's Medical Centre, was transported to the hospital in the early hours of August 8 by her mother.

Doctors informed her that she could had died of a heart attack while sleeping. Billy's quick efforts, she now believes, saved her life. "I was surprised, but I went to bed and felt fine. I'd even been out with the dogs and I hadn't felt ill or had any pains," she says in an interview with SWNS. "I awoke in the early hours covered in sweat and unable to move." Billy was perched on my chest, meowing loudly in my ear canal. He was meowing a lot. He doesn't generally do that; his life revolves on sleeping all day and all night. Doctors determined that one of Felstead's arteries had become obstructed, resulting in the heart attack. A balloon was inserted into her artery, and she will now be on medicine for the remainder of her life. It's a tiny price to pay for a scenario that could have killed her.

Stories about pets going the extra mile to assist their owner in distress support prior study indicating that dogs will rescue their people in extreme situations. Arizona State University researchers studied the reactions of 60 companion dogs while their owners appeared to be trapped inside a "locked" box in 2020. None of the dogs in the study had any prior experience or training in human rescue. While the owners were inside, they summoned their dogs for assistance. Researchers found it impressive that one-third of the dogs ended up rescuing their owners. They claim the dogs' heroic deeds were noteworthy for two reasons. The dogs certainly wanted to help their owners, and they were also bright enough to figure out what measures needed to be made (opening the box's door).

Two further control tests in the study demonstrated that most dogs responded physiologically to the cries of their distraught owners. "What's fascinating about this study," concluded co-author Clive Wynne, "is that it demonstrates that dogs truly care about their people." Even in the absence of training, many dogs will attempt to rescue people who look to be in difficulty — and even if they fail, we can see how sad they are. The control test results show that dogs who fail to rescue their owners are unable to understand what to do — it's not because they don't care about their owners.

But does this imply that dogs are more attached to their owners than cats? According to a 2019 Oregon State University study, cats love their human caregivers exactly as much as any dog or child! Researchers found for the first time that cats had the same attachment types as dogs. A group of cats took part in a "secure base test," in which each cat spent two minutes in a room with their owner, two minutes alone, and then two minutes reunited with their owner. The test's ultimate purpose was to categorize each feline's attachment tendencies as secure or insecure. Similar attachment tests have previously been performed on both canines and human neonates. Infants and animals who exhibit "secure attachment" often trust their caregivers. They look to them for safety and even assume that if they are separated, their caregiver will return to them.

When the owners of the cats returned for the two-minute reunion phase, researchers observed how each creature reacted. If the cat showed no signs of worry and was not surprised when their person returned, the animal was labeled as having a "secure attachment." In contrast, if the cat reacted to seeing their owner by showing signs of tension (tail twitching, lip licking) and either avoiding or initiating (jumping on their lap) contact, the cat was classed as having "insecure attachment." These tests were carried out on both kittens and adult cats. Following that, a team of behavioral experts watched the experiment's recorded film and matched each cat's actions to criteria commonly used to determine attachment levels in dogs and human infants.

Of the 70 kittens that could be classified, 64.3% were classed as firmly connected and 35.7% were classified as insecurely attached. The researchers also wanted to see if socializing training had any effect on these data, so they put the kittens in a six-week training session and re-ran the experiment. Surprisingly, there were no significant differences in the outcomes of the second experiment.

Furthermore, the researchers were shocked by how closely the ratios of secure and insecure relationships in feline groupings paralleled findings in a human neonate community. 65% of infants were found to be securely connected, only 8% less than the feline group.

There are numerous established benefits to owning a pet, and having them there to aid you in extreme situations is another.

Question: 1 of 30

According to Extract A, what is one characteristic that sets cats apart from dogs?

- **A.** Their ability to communicate through vocalizations.
- **B.** Their independent nature and preference for personal space.
- **C.** Their instinctive hunting skills.
- **D.** Their reliance on scent marking for communication.

Question: 2 of 30

According to Extract A, what is the importance of providing an enriched environment for cats?

- **A.** It helps maintain their physical health.
- **B.** It fosters a stronger bond with their human companions.
- **C.** It allows them to engage in natural behaviors and mental stimulation.
- **D.** It ensures they have a clean litter box and fresh water.

Question: 3 of 30

According to Extract A, what is a responsible practice of cat ownership?

- **A.** Regular playtime and interactive sessions.
- **B.** Feeding them a balanced diet and providing veterinary care.
- **C.** Ensuring access to clean water and a clean litter box.
- **D.** All of the above.

Question: 4 of 30

According to Extract B, how did the cat Billy help save his owner's life?

- **A.** He called out for help when the owner suffered a heart attack.
- **B.** He woke up the owner by pounding his paws on her chest.
- **C.** He ran to get the owner's mother for assistance.
- **D.** He alerted the owner by jumping from the turret.

Question: 5 of 30

According to Extract B, what did the 2020 study by Arizona State University researchers reveal about dog's reactions in a distressing situation?

- **A.** Dogs were trained to rescue their owners in a locked box.
- **B.** Dogs showed a physiological response to their distressed owner's cries.
- **C.** Dogs were indifferent to their owner's calls for help.
- **D.** Dogs lacked the intelligence to figure out how to open a locked box.

Question: 6 of 30

According to Extract B, what did the study from Oregon State University conclude about the attachment styles of cats?

- **A.** Cats are less attached to their owners compared to dogs.
- **B.** Cats only exhibit attachment behaviors when they require food or attention.
- **C.** Cats show similar attachment styles to dogs and human infants.
- **D.** Cats are more likely to display insecure attachments than dogs.

Direction for Questions: 7 to 9

Read the text below and answer the questions.

Nature's Nutrient Powerhouse: The Versatile Coconut

The coconut is a tropical gem that has been adored for ages for its cooling water and creamy flesh, as well as its myriad health benefits and versatile uses. The coconut tree (Cocos nucifera), sometimes known as the "tree of life" in various cultures, provides a variety of items ranging from nutritious food to natural beauty cures. Let's take a look at the intriguing world of coconuts and see why they're regarded as a nutritious powerhouse.

Coconuts are well-known for their distinctive shape, which consists of a strong outer husk around a dark, fibrous shell. The coconut meat, which is rich, creamy, and nutrient-dense, is found inside the shell. Coconuts are highly prized in tropical places where they are indigenous, as they provide food, water, and other essential resources to local inhabitants.

Coconut flesh is high in beneficial fats, such as medium-chain triglycerides (MCTs), which are readily absorbed and transformed into energy. MCTs have been linked to a variety of health benefits, including better cognitive function, higher satiety, and faster metabolism. Coconuts are also high in fiber, which helps with digestion and promotes gut health. They also include essential minerals such as potassium, manganese, and copper.

The moisturizing and thirst-quenching coconut water is one of the most delightful properties of coconuts. This natural beverage is encased within young green coconuts and is a natural supply of electrolytes such as potassium, magnesium, and salt. Many athletes and fitness enthusiasts choose coconut water to sugary sports beverages because it helps replace fluids and minerals lost during physical activity.

Coconut is a versatile ingredient that adds taste and texture to a variety of dishes. Coconut milk is a common component in many Asian and Caribbean dishes. It is created by blending grated coconut meat with water. It gives curries, soups, and desserts a creamy richness. Coconut oil, made from the meat of mature coconuts, is commonly used in cooking, baking, and even as a vegan substitute for butter. Coconut oil has grown in popularity due to its distinct aroma and silky texture, as well as its possible health advantages and culinary versatility.

Coconut-based beauty and health products have also made their way into the market. Coconut oil is a popular skincare product because of its hydrating characteristics and ability to nurture the skin. It's a natural moisturizer, makeup remover, and hair conditioner all in one. Coconut oil is also used in traditional oil-pulling methods, which include swishing oil around in the mouth to promote oral health and fresh breath.

Coconuts are also environmentally beneficial, in addition to their other benefits. Coconut trees flourish in a variety of habitats and require little upkeep, making them a sustainable crop. They are particularly well-known for their adaptability, as various elements of the tree, such as the trunk, leaves, and shells, can be used for a variety of uses, such as construction, fuel, and crafts. Furthermore, coconuts are biodegradable, which contributes to a lower environmental effect when compared to synthetic materials.

Finally, the coconut truly merits its image as a nutrient-dense and adaptable natural resource. The coconut continues to fascinate our taste buds, increase our well-being, and contribute to sustainable practices, thanks to its nutritious characteristics and hydrating water, as well as its culinary delights and beauty benefits. So, the next time you eat a great coconut-based food or use coconut oil in your beauty routine, relish it.

Question: 7 of 30

What are medium-chain triglycerides (MCTs) found in coconuts known for?

- **A.** Supporting brain function
- **B.** Aiding in digestion
- **C.** Boosting metabolism
- **D.** All of the above

Question: 8 of 30

How is coconut oil commonly used in skincare?

- **A.** Hair conditioner
- **B.** Moisturizer
- **C.** Makeup remover
- **D.** All of the above

Question: 9 of 30

What makes coconuts an environmentally friendly crop?

- **A.** They require minimal maintenance.
- **B.** Different parts of the tree can be utilized for various purposes.
- **C.** They contribute to reduced environmental impact.
- **D.** All of the above

Direction for Questions: 10 to 14

Read the poem below then answer the questions.

The Tide Rises, the Tide Falls
by **Henry Wadsworth Longfellow**

The tide rises, the tide falls,
The twilight darkens, the curlew calls;
Along the sea-sands damp and brown
The traveller hastens toward the town,
And the tide rises, the tide falls.

Darkness settles on roofs and walls,
But the sea, the sea in the darkness calls;
The little waves, with their soft, white hands,
Efface the footprints in the sands,
And the tide rises, the tide falls.

The morning breaks; the steeds in their stalls
Stamp and neigh, as the hostler calls;
The day returns, but nevermore
Returns the traveller to the shore,
And the tide rises, the tide falls.

Question: 10 of 30

In the second stanza, what do the "soft, white hands" of the waves symbolize?

- **A.** The welcoming embrace of the sea
- **B.** The traveler's footprints in the sand
- **C.** The darkness of the evening
- **D.** The passage of time

Question: 11 of 30

What is the deeper meaning or theme of the poem as reflected in the repeated line "And the tide rises, the tide falls"?

- **A.** The cyclical nature of life and death
- **B.** The unpredictability of nature
- **C.** The eternal beauty of the sea
- **D.** The traveler's persistence in his journey

Question: 12 of 30

How does the poet create a sense of melancholy in the poem?

- **A.** Through descriptions of joyful moments
- **B.** Through the use of vivid colors
- **C.** Through the repetition of the tide's movement
- **D.** Through the traveler's inability to return

Question: 13 of 30

What effect does the repetition of the line "And the tide rises, the tide falls" have **on the poem's structure**?

- **A.** It adds a sense of rhythm and continuity.
- **B.** It creates a disjointed and chaotic structure.
- **C.** It emphasizes the beauty of sea.
- **D.** It signifies the changing of seasons.

Question: 14 of 30

Which stanza of the poem portrays a sense of renewal or rebirth?

- **A.** The first stanza
- **B.** The second stanza
- **C.** The third stanza
- **D.** None of the stanzas

Direction for Questions: 15 to 18

Read the extracts and then answer the questions.

Extract A:
Botany, the scientific study of plants, is a fascinating field that explores the morphology, physiology, ecology, and evolution of plant life. Botanists delve into the intricate structures and functions of plants, examining their growth, reproduction, and interactions with the environment. This field is essential for understanding the role of plants in sustaining ecosystems, agriculture, and medicine. Botany also contributes to conservation efforts by studying endangered plant species and their habitats. Through research and experimentation, botanists uncover the secrets of plants, unlocking valuable knowledge that can be applied to various fields, such as agriculture, forestry, and pharmaceuticals.

Extract B:
Zoology, the scientific study of animals, is a diverse and captivating field that encompasses the exploration of animal behavior, physiology, and evolution. Zoologists investigate the intricacies of animal life, studying their anatomy, habitats, and interactions with other species. This field allows us to understand the diverse range of animal species on our planet, from microscopic organisms to complex ecosystems. Zoology plays a significant role in conservation efforts, as scientists work to protect endangered species and their habitats. Furthermore, zoologists contribute to the fields of medicine and agriculture by studying animal diseases and their impact on human health and food production.

Extract C:
Anatomy, the study of the structure and organization of living organisms, is a fundamental field in the biomedical sciences. It explores the intricate details of the human body, from the macroscopic level down to the cellular and molecular level. Anatomists examine the relationships between different body parts, organs, and systems, unraveling the complexities of human form and function. This field is crucial for medical professionals, providing the foundation for diagnosing and treating diseases. It also plays a significant role in surgical procedures and medical research. By understanding the intricacies of anatomy, scientists and healthcare professionals can improve human health and enhance medical advancements.

Extract D:
Horticulture, the science and art of cultivating plants, is a field that encompasses the study of plant cultivation, propagation, and management. Horticulturists explore various aspects of plant life, from growing and maintaining ornamental plants to cultivating crops for food production. This field combines scientific knowledge with practical skills, such as plant breeding, pest management, and landscape design. Horticulture plays a critical role in agriculture, ensuring sustainable food production and the development of new plant varieties. It also contributes to the aesthetic beauty of our surroundings, as horticulturists design and maintain gardens, parks, and green spaces. Through horticulture, we can harness the potential of plants for practical and aesthetic purposes.

Question: 15 of 30

Which of the extracts refers to the science and art of cultivating plants?

A. Extract A
B. Extract B
C. Extract C
D. Extract D

Question: 16 of 30

Which of the extracts refers to a field that encompasses the exploration of animal behavior?

A. Extract A
B. Extract B
C. Extract C
D. Extract D

Question: 17 of 30

Which of the extracts refers to the understanding of the role of plants in sustaining ecosystems?

A. Extract A
B. Extract B
C. Extract C
D. Extract D

Question: 18 of 30

Which of the extracts talks about the conservation of endangered species?

A. Extract A
B. Extract B
C. Extract C
D. Extract D

Direction for Questions: 19 to 25

Seven sentences have been removed from the text. Choose from the sentences (A – H) the one which fits each gap (1 – 7). There is one extra sentence which you do not need to use.

The Hidden World of Tiny Life

1.... They exist in vast numbers and play a crucial role in the world around us. From the depths of the ocean to the soil beneath our feet, microorganisms are present everywhere, silently shaping our planet.

One of the most fascinating aspects of microorganisms is their incredible diversity. **2...** Despite their small size, microorganisms have a profound impact on the environment and on other living organisms.

Microorganisms are essential for maintaining the balance of ecosystems. **3...** Without microorganisms, dead plants and animals would accumulate, leading to a buildup of waste and a lack of essential nutrients for other organisms.

4... They are used in various biotechnological processes, such as the production of antibiotics, enzymes, and biofuels. Microorganisms also play a crucial role in food production, as they are involved in processes like fermentation, which is used to make bread, cheese, and yogurt.

Microorganisms are not only beneficial but can also be harmful to humans. **5...** However, it is important to note that only a small fraction of microorganisms are harmful, and the majority are either beneficial or neutral.

6.... They have been used as model organisms to study biological processes and have contributed to our understanding of genetics, evolution, and immunology. Microorganisms have also been used in genetic engineering and biotechnology, enabling scientists to manipulate their genetic material for various purposes.

In recent years, the study of microorganisms has gained even more attention due to the emerging field of microbiomics. **7....** Research in this area has revealed the crucial role that microorganisms play in maintaining our health and well-being.

In conclusion, microorganisms are a vital and diverse group of organisms that have a significant impact on our planet. They are responsible for maintaining the balance of ecosystems, have various industrial applications, and have contributed to scientific advancements. Understanding and appreciating the hidden world of microorganisms can help us better appreciate the intricate web of life that surrounds us.

A. ***They come in various shapes, sizes, and forms, ranging from bacteria and fungi to viruses and protozoa.***
B. ***They play a crucial role in nutrient cycling, breaking down organic matter and releasing nutrients into the environment.***
C. ***Some microorganisms cause diseases, such as bacteria that cause infections and viruses that cause illnesses like the flu.***
D. ***Microorganisms have also been instrumental in scientific research and advancements.***
E. ***In addition to their ecological importance, microorganisms have numerous industrial applications.***
F. ***They are responsible for maintaining the balance of ecosystems.***
G. ***Microorganisms, also known as microbes, are living organisms that are too small to be seen with the naked eye.***
H. ***This field focuses on studying the communities of microorganisms that live in and on our bodies, known as the human microbiome.***

Question: 19 of 30

Which one fits in (1) ?

- **A.** Sentence A
- **B.** Sentence B
- **C.** Sentence G
- **D.** Sentence H

Question: 20 of 30

Which one fits in (2) ?

- **A.** Sentence A
- **B.** Sentence B
- **C.** Sentence C
- **D.** Sentence D

Question: 21 of 30

Which one fits in (3) ?

- **A.** Sentence A
- **B.** Sentence D
- **C.** Sentence B
- **D.** Sentence E

Question: 22 of 30

Which one fits in (4) ?

- **A.** Sentence A
- **B.** Sentence B
- **C.** Sentence H
- **D.** Sentence E

Question: 23 of 30

Which one fits in (5) ?

- A. Sentence A
- B. Sentence B
- C. Sentence C
- D. Sentence D

Question: 24 of 30

Which one fits in (6) ?

- A. Sentence A
- B. Sentence B
- C. Sentence C
- D. Sentence D

Question: 25 of 30

Which one fits in (7) ?

- A. Sentence A
- B. Sentence G
- C. Sentence H
- D. Sentence C

Direction for Questions: 26 to 30

Read the poem then answer the questions.

On His Blindness
***by* John Milton**

When I consider how my light is spent,
Ere half my days, in this dark world and wide,
And that one Talent which is death to hide
Lodged with me useless, though my Soul more bent
To serve therewith my Maker, and present
My true account, lest he returning chide;
"Doth God exact day-labour, light denied?"
I fondly ask. But patience, to prevent
That murmur, soon replies, "God doth not need
Either man's work or his own gifts; who best
Bear his mild yoke, they serve him best. His state
Is Kingly. Thousands at his bidding speed
And post o'er Land and Ocean without rest:
They also serve who only stand and wait."

Question: 26 of 30

What is the tone of the poem's opening lines?

- A. Joyful
- B. Despairing
- C. Indifferent
- D. Ironic

Question: 27 of 30

In the poem, what does the speaker consider a "Talent"?

- **A.** His ability to write poetry
- **B.** His physical strength
- **C.** His knowledge of the world
- **D.** His wealth and possessions

Question: 28 of 30

What is the deeper meaning or message conveyed by the poem?

- **A.** The importance of worldly success
- **B.** The inevitability of physical decline
- **C.** The value of patience and acceptance in serving God
- **D.** The futility of serving God without expecting rewards

Question: 29 of 30

Which phrase from the poem suggests that God does not require human efforts or offerings?

- **A.** "my light is spent"
- **B.** "God doth not need"
- **C.** "they serve him best"
- **D.** "his mild yoke"

Question: 30 of 30

What does the speaker mean by "his mild yoke" in the poem?

- **A.** God's gentle guidance
- **B.** God's heavy burden
- **C.** God's strict rules
- **D.** Our material wants

Preptive Prepare Better

Series D Mathematical Reasoning Trial Test

Name: ______________________

Date: ______________________

Question: 1 of 35

I am a prime number between 10 and 40. I am a factor of 111.

What number am I ?

A. 35
B. 36
C. 37
D. 38

Question: 2 of 35

Noah bought 231 eggs. Out of this, he sold $^{9}/_{11}$ of eggs and $^{2}/_{7}$ of the **remaining eggs** were broken. How many eggs will Noah have at the end?

A. 30
B. 42
C. 36
D. 41

Question: 3 of 35

At Fresh Bread and Cakes bakery, the price of one fresh loaf of bread is $1. However, if there are leftovers, each of these loaves is sold $0.40 cheaper on the next day. The table below shows the number of loaves baked and sold over two days.

On August 2nd, all loaves were sold. Find the total earnings for the two days.

Days	Number of loaves baked	Total number sold
August 1st	90	62
August 2nd	90	118

A. $168.80
B. $170.00
C. $124.30
D. $190.00

Question: 4 of 35

John and Peter shared some marbles in the ratio of 7:5. John had 24 more marbles than Peter.
How many marbles did Peter have?

A. 12
B. 60
C. 84
D. 144

Question: 5 of 35

Sam spent $^{1}/_{4}$ of his money at recess. He spent $^{1}/_{6}$ of his **remaining money** on lunch. What fraction of his money was left in the end?

- **A.** $^{1}/_{8}$
- **B.** $^{5}/_{8}$
- **C.** $^{5}/_{12}$
- **D.** $^{7}/_{12}$

Question: 6 of 35

A rope of length 72 cm was cut into three pieces. The first piece was three times as long as the second piece. The second piece was twice as long as the third piece.

How long was the second piece?

- **A.** 12 cm
- **B.** 16 cm
- **C.** 18 cm
- **D.** 24 cm

Question: 7 of 35

In a garden there is a row of trees, one tree is fifth from either ends of this row.

How many trees are there in the row?

- **A.** 5
- **B.** 8
- **C.** 10
- **D.** 9

Question: 8 of 35

Linda was given 100 minutes to complete a test. She took $\frac{3}{4}$ of the time to complete the test and she stopped writing after that. She started doing the test at 14:50.

At what time did she stop writing?

- **A.** 4:00 PM
- **B.** 4:05 PM
- **C.** 5:00 PM
- **D.** 6:05 PM

Question: 9 of 35

A photo frame measures 15 centimeters by 10 centimeters. A photograph is placed in it, leaving a border of 2 centimeters wide all around it.

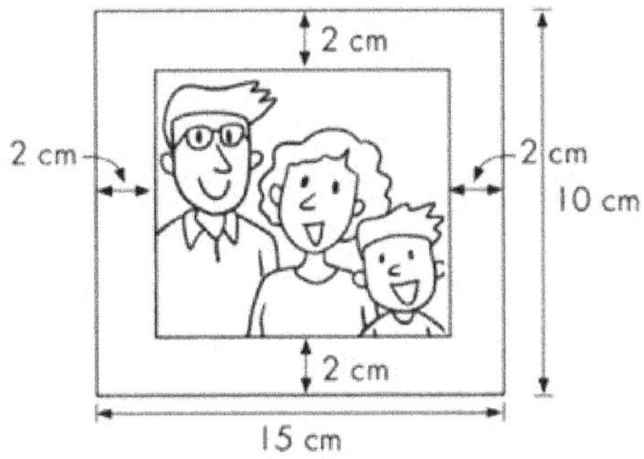

What is the area of the border?

A. 80 sq cm

B. 84 sq cm

C. 100 sq cm

D. 160 sq cm

Question: 10 of 35

Grandpa had fewer than 20 finger puppets. If he gave each of his grandchildren 3 finger puppets, he would have 2 finger puppets left. If he gave each grandchild 4 finger puppets, he would be short of 3 finger puppets.

How many finger puppets did he have?

A. 10

B. 11

C. 17

D. 20

Question: 11 of 35

5 notebooks, 2 pencils and 2 markers cost \$10. 1 pencil and 1 marker cost \$2.
How much does 1 notebook cost?

A. \$6.20

B. \$1.20

C. \$4.00

D. \$10.00

Question: 12 of 35

At a fruit shop, mangoes are sold at 5 for \$9 while apples are sold at 7 for \$8. Mr Rao bought an equal number of mangoes and apples. He spent \$92 more on the mangoes.
How many apples did Mr Rao buy?

A. 150

B. 5

C. 140

D. 144

Question: 13 of 35

The table below shows the opening hours of a clinic in a day.

Opening Hours
8.30 a.m. - 12.45 p.m.
2.00 p.m. - 4.30 p.m.
7.00 p.m. - 8.30 p.m.

How long is the clinic open per day ? Leave your answer in hours and minutes.

- **A.** 10h 30min
- **B.** 2h 10min
- **C.** 6h 14min
- **D.** 8h 15min

Question: 14 of 35

A rectangular tank measuring 150 cm by 70 cm by 85 cm high was $\frac{2}{5}$ filled with water. Mrs Sam poured 12 pails of water, each containing the same volume of water, into the tank. In the end, there was 615 L of water in the tank. What was the volume of water in each pail?
Give your answer in litres.

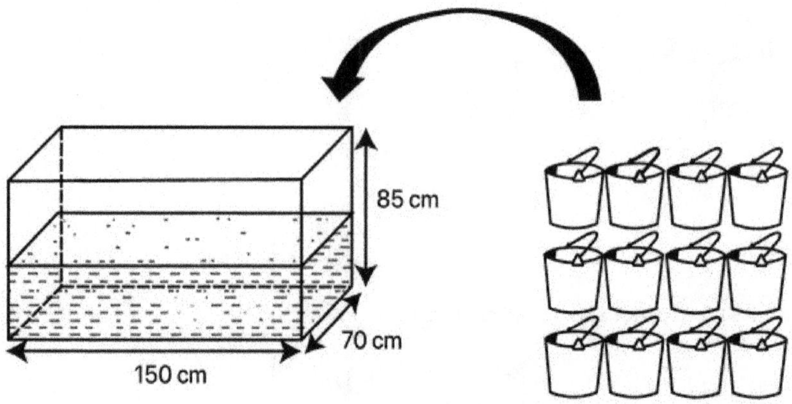

- **A.** 21.5L
- **B.** 23.6L
- **C.** 33.8L
- **D.** 22.9L

Question: 15 of 35

Ravi has some marbles. If he gives 7 marbles to each of his friends, he will have 4 marbles left. If he gives each of his friends 9 marbles, he will need another 2 more marbles.
How many marbles does he have?

- **A.** 39
- **B.** 25
- **C.** 26
- **D.** 90

Question: 16 of 35

The mass of a glass bottle filled with 10 identical erasers is 1200 g. The mass of the same bottle when filled with 6 such erasers is 1040 g. What is the mass of each eraser?

- **A.** 44g
- **B.** 49g
- **C.** 45g
- **D.** 40g

Question: 17 of 35

Mr Tan made some fish balls. He sold $\frac{3}{4}$ of the fish balls in the morning and $\frac{1}{3}$ of the remaining fish balls in the afternoon. He made another 506 fish balls and had twice the number of fish balls he made at first.
How many fish balls did he make at first?

- **A.** 276
- **B.** 65
- **C.** 287
- **D.** 90

Question: 18 of 35

A total of $2,500 is shared between 3 pupils, Zay, Jack, and Rose. Rose gets $625. Zay gets 20% more money than Rose. Calculate the money that Jack receives.

- **A.** $ 1,125
- **B.** $ 1,115
- **C.** $ 1,525
- **D.** $ 1,365

Question: 19 of 35

The table shows the symbols and their values.

Symbol	Value
	35
	65
	37
	102
	65

Which of the followings is correct?

- **A.**  -  = 
- **B.**  -  = 
- **C.**  -  = 
- **D.**  -  = 

Question: 20 of 35

The bar chart below shows the number of people who likes a variety of fruits.

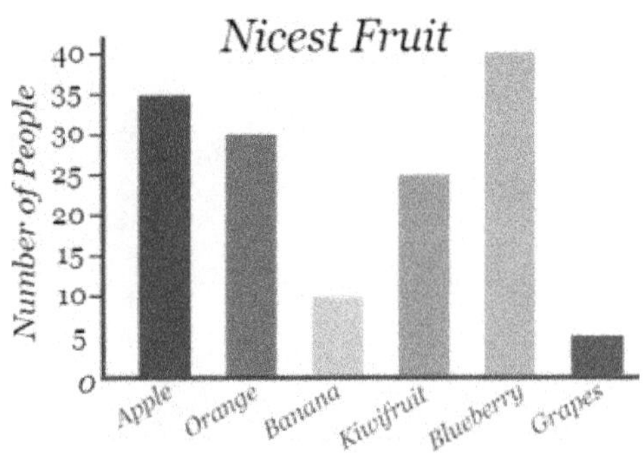

Which of the following is a correct statement?

- **A.** The number of people who like bananas and apples is 50.
- **B.** The number of people who like grapes and oranges is 45.
- **C.** The number of people who like kiwifruit and blueberry is 75.
- **D.** The number of people who like orange and blueberry is 70.

Question: 21 of 35

The table shows 4 colors of pens in a box.

Color	Pens
Purple	35%
Red	25%
Blue	
Black	10%

There were 7,000 pens in a box. Half of the blue is sold. How many **blue pens are remaining unsold** ?

A. 1,527

B. 1,505

C. 1,023

D. 1,050

Question: 22 of 35

The pictures show a surface area of a shape.

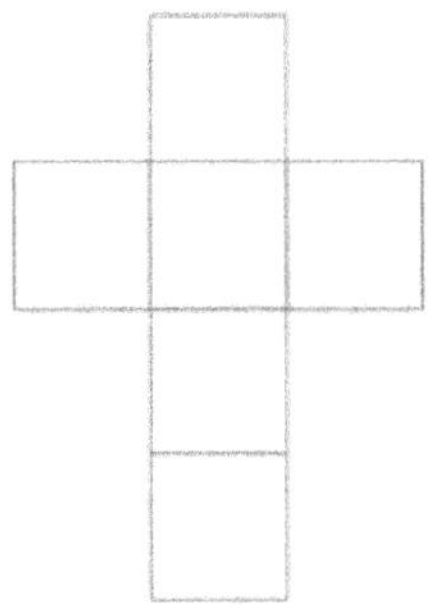

What shape will appear when the surface area got folded?

A. Cube

B. Pyramid

C. Cone

D. Cylinder

Question: 23 of 35

In a classroom activity, 16 pupils stood in a straight row with equal distances between each other. The distance between the 5th pupil and the 10th pupil was 11.95 m.
What was the distance between the first pupil and the last pupil?

A. 38.35 m

B. 36 m

C. 35.85 m

D. 37.5 m

Question: 24 of 35

The table below shows the charges of an automated bicycle rental system in city.

Rental charges	
For the first hour	$1.50
For every subsequent $\frac{1}{2}$ hour or part thereof	$0.50

How much must Michelle pay to rent the bicycle from 1.30 p.m. to 4.15 p.m.?

- **A.** $3.50
- **B.** $3.00
- **C.** $4.50
- **D.** $5.00

Question: 25 of 35

Ali, Ben and Carson shared some sweets. The ratio of the number of sweets Ali had to the total number of sweets Ben and Carson had is 3 : 8. Ali had the same number of sweets as Ben and 36 fewer sweets than Carson.
What is the ratio of the number of sweets Ali had to the number of sweets Ben had to the number of sweets Carson had?

- **A.** 3:3:8
- **B.** 3:5:3
- **C.** 3:3:3
- **D.** 3:3:5

Question: 26 of 35

The average test score of a group of students was 72 marks. After three students who scored an average of 85 marks left the group, the average score of the remaining students became 69 marks. How many students were there in the group at first?

- **A.** 76
- **B.** 156
- **C.** 36
- **D.** 16

Question: 27 of 35

Which of the following statements correctly describes the number of edges, vertices, and faces of a cube?

- **A.** 6 edges, 8 vertices, 6 faces
- **B.** 8 edges, 6 vertices, 12 faces
- **C.** 12 edges, 8 vertices, 6 faces
- **D.** 4 edges, 6 vertices, 4 faces

Question: 28 of 35

Table below shows the number of phones owned by students in school S.

Phone	**Number of phones**
Samsung	850
Huawei	60% of Samsung
Apple	561 more than Huawei

Calculate the total number of phones owned by the students.

- **A.** 3,410
- **B.** 8,240
- **C.** 2,431
- **D.** 2,410

Question: 29 of 35

When Ahmed walked through a shopping mall, he recorded the number of customers in each shop. The data is displayed in a dot plot.

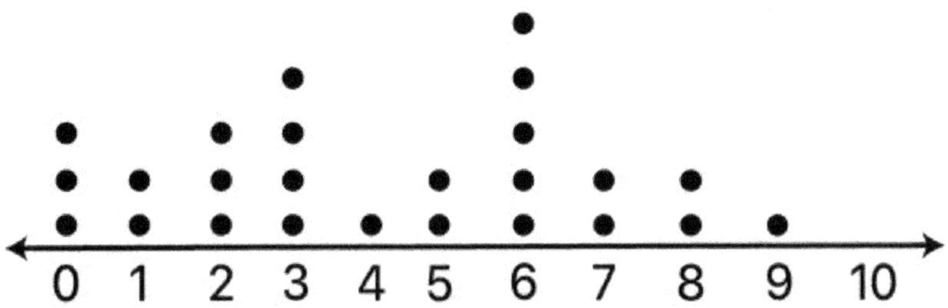

What percentage of shops had at least 5 customers?

- **A.** 25%
- **B.** 48%
- **C.** 33%
- **D.** 44%

Question: 30 of 35

In a mixture of 60 litres, the ratio of milk and water is 2:1. If the ratio is to be 1:2, then the quantity of water to be further added is ..

- **A.** 20 L
- **B.** 30 L
- **C.** 50 L
- **D.** 60 L

Question: 31 of 35

Sofia and her friends ordered 3 large pan pizza. Each pizza was cut into 8 pieces.

Iris ate 5 pieces.
Luna ate $^1/_6{}^{th}$ of the total.
Eloise ate $^1/_3{}^{rd}$ of the total, and Sofia ate the remaining.
Who ate most?

- **A.** Iris
- **B.** Luna
- **C.** Eloise
- **D.** Sofia

Question: 32 of 35

This chart denotes favourite ice-cream falvours of people.

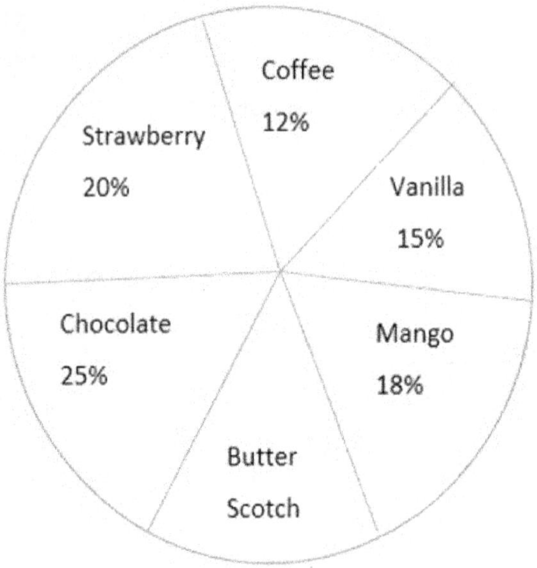

If 735 people like vanilla flavour, how many people like mango flavour?

- **A.** 752
- **B.** 798
- **C.** 844
- **D.** 882

Question: 33 of 35

$\angle X = 52°$. $\angle Y$ is twice that of $\angle X$. Find $\angle Z$. (Figure not drawn to scale.)

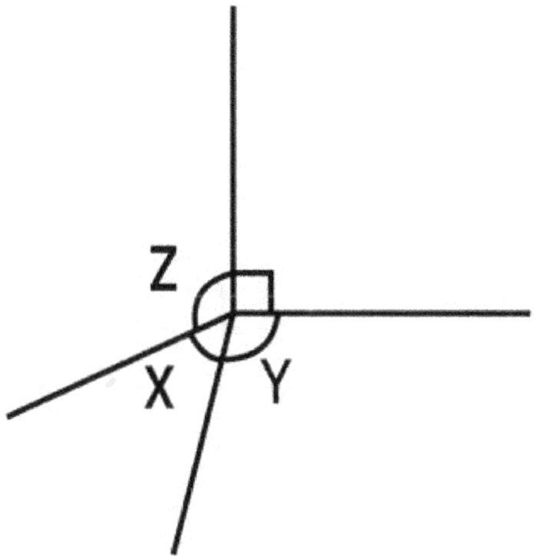

A. 114°

B. 104°

C. 124°

D. 134°

Question: 34 of 35

Cheryl earned a fixed income each month. In September, Cheryl spent some money and saved the rest. In October, the amount she spent increased by 20% and she saved 40% less compared to the previous month.

What fraction of Cheryl's income in September did she spend?

A. $^1/_3$

B. $^2/_3$

C. $^4/_3$

D. $^3/_2$

Question: 35 of 35

In the figure shown below, QRUV is a square. Line QS is 15 cm and line WU is 35 cm. What is the perimeter of rectangle PSTW?

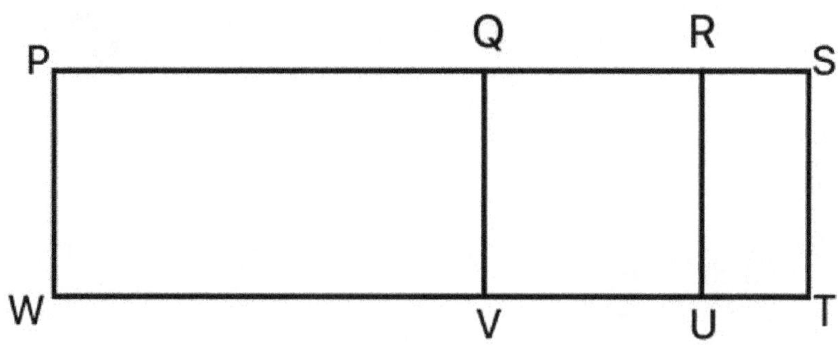

A. 100 cm

B. 120 cm

C. 50 cm

D. 110 cm

Preptive Prepare Better

Series D Thinking Skill Trial Test

Name: ______________________________

Date: ______________________________

Question: 1 of 40

Dilan: "To have a greater chance of winning on Archery, you must practice for at least 30 minutes, four times a week."

Steve: "I do more than 30 minutes every day. I am so sure to win the Archery."

Which of these statements show the mistake that Steve has made?

- **A.** The amount of practice time doesn't have any effect on the chance of winning.
- **B.** More practice time can not guarantee a greater chance of winning.
- **C.** Less than 30 minutes of practice four times a week will lead to winning.
- **D.** Less than 30 minutes of practice three times a week will lead to winning.

Question: 2 of 40

Mrs. Lee gives scores in lecture and laboratory to each student in his Chemistry class. These two scores are then added to give an overall score in Chemistry for the term (for example, a student with score 50 in lecture and 30 in laboratory would have an overall score of 80 in Chemistry.)

Last term Ace and Ben got the same overall score in Chemistry.

Ace: "If our scores in Lecture were the same, then our score in Laboratory must have been the same too."

Ben: "If our scores in Laboratory were different from each other, then our score in Lecture must have been different too."

If the information is true, whose reasoning is correct?

- **A.** Ace only
- **B.** Ben only
- **C.** Both Ace and Ben
- **D.** Neither Ace nor Ben

Question: 3 of 40

According to a survey of color preference: everyone who like pink liked red. Everyone who like red liked orange but no one who like red liked grey.
Kim, Loisa, Liza and Jane all took part in the survey.

Based on the information, which of the following must be true?

- **A.** If Kim like orange, she also likes red.
- **B.** If Loisa doesn't like grey, she doesn't like orange.
- **C.** If Jane likes pink, she does not like grey.
- **D.** If Liza doesn't like pink, she doesn't like red.

Question: 4 of 40

Kacey and her friends like fruits very much. Kacey loves oranges, bananas, and mangoes while Lily loves apples and oranges. Rachel doesn't like bananas and apples, but she loves all the other fruits. Which of the following is correct regarding the above situation?

- **A.** Rachel likes all fruits that Kacey likes
- **B.** Kacey doesn't like any fruit that Lily likes
- **C.** Lily likes all fruits that Rachel likes
- **D.** Lily likes some fruits that Rachel likes

Question: 5 of 40

Adam is a software developer. Usually, it takes 6 days to complete software development and if the client needs to get the work done in less than 6 days, the price will be inverse proportionally increased. e.g., if the client needs him to finish the software in 3 days, the price will be doubled.
If it usually costs $80 for a specific software development and a client needs Adam to finish it within 2 days, how much will it cost client?

- **A.** 160
- **B.** 240
- **C.** 40
- **D.** 120

Question: 6 of 40

Jill has cricket practices **every** evening.
If she missed practices four days in a row last week, on which day did she definitely miss the practices?

Note : A week is Monday to Sunday

- **A.** Saturday
- **B.** Monday
- **C.** Thursday
- **D.** Wednesday

Question: 7 of 40

These are representations of three English words.

2465, 1465, 2769

What can the words be ?

- **A.** Ring, Sing, Roll
- **B.** Reel, Wheel, Real
- **C.** Near, Node, Null
- **D.** Rank, Bank, Ring

Question: 8 of 40

Following tables show the marks of students on two science exams.

Exam 01

Hannah	60
Robin	55
Kathy	65
David	75

Exam 02

Robin	79
David	78
Hannah	82
Kathy	82

If the total marks of two papers of a student **who has made the least improvement** (from Exam 01 to Exam 02) is less than 135, he or she has to face a viva.

Who will have to face a viva?

- **A.** Hannah
- **B.** Kathy
- **C.** David
- **D.** No one

Question: 9 of 40

Lily's cookery class teacher instructed to use : 2 eggs, 4 ½ cups of milk, 100 grams of sugar, and a teaspoon of salt to make the previously taught recipe for 2 people.
Lily wanted to make it this morning, and she needed 22 cups of milk. How many people are there at Lily's apartment?

A. 5
B. 6
C. 7
D. 8

Question: 10 of 40

Barney went to buy a watch, and this is what he found out about the prices.
Piaget watches are expensive, but not like Rolex watches. Blancpain watches were more expensive than Longines watches, but not like Piaget ones. Omega watches are cheaper than Tudor watches, and are the cheapest of the lot.
Which brand has the 3rd cheapest watches?

A. Tudor
B. Rolex
C. Piaget
D. Longines

Question: 11 of 40

If it is Monday, George would attend cricket practices after school.
And also, if George doesn't attend practices, this means he is attending urgent tutoring class in the evening.

It is Monday and George has no tutoring class. So :

A. He will attend practices
B. He won't attend practices
C. He will go home after the school
D. Cannot decide whether he attends the practices or not

Question: 12 of 40

Weather is very important in farming. Every farmer knows well about the weather.

Kevin: "My uncle knows well about the weather."

So...

A. Kevin's uncle is a farmer
B. Kevin's uncle is not a farmer
C. Kevin's uncle may be a farmer
D. None of the above is correct

Question: 13 of 40

Principal : "The students of each class who got the highest marks for Mathematics will be given a prize by next week."

It's been two months since principal's announcement, but Daniel didn't get a prize. That means,

- **A.** There was another student who got the lower marks
- **B.** Some students got higher marks than him in his class
- **C.** Daniel misses everything
- **D.** None of the above is correct

Question: 14 of 40

Please refer to the statements below. For the purpose of this question, they are correct even if they contradict commonly known facts.

None of the vehicles are eco-friendly.
Some vehicles are electric vehicles.

Which of the following can be a conclusion ?

- **A.** Only non-electric vehicles are not ecofriendly
- **B.** Some vehicles are ecofriendly
- **C.** Some electric vehicles are not ecofriendly
- **D.** All the electric vehicles are not ecofriendly

Question: 15 of 40

Harry: "If Emily goes to the party, she will be late to work tomorrow."

Rupert: "I think that won't happen."

What does Rupert think?

- **A.** Emily won't go to the party
- **B.** Emily will go to the party, and she won't be late to work
- **C.** Emily won't either go to the party or be late to work
- **D.** None of the above

Question: 16 of 40

Jenny and Joey were waiting at bus stop when they saw Emma was approaching the bus stop.

Jenney: "If it rains, Emma will miss the bus."

Joey: "Oh no. She gets angry whenever she misses the bus."

It is heavily raining since morning. Which is a next possible situation ?

- **A.** Emma is angry now
- **B.** Maybe Emma is angry now
- **C.** Emma is not angry
- **D.** No conclusion

Question: 17 of 40

Whenever Eunice is stressed, she takes a warm bath to relax. When she takes a warm bath, she eats her comfort food which is ice cream.

Teresa: "Eunice is stressed. She must be taking a warm bath to relax."

Jackie: "If Eunice took a warm bath then she must be eating an ice cream."

Whose reasoning is correct?

- **A.** Teresa
- **B.** Jackie
- **C.** Teresa and Jackie
- **D.** Neither Teresa nor Jackie

Question: 18 of 40

Carlo said that if Kim does not study hard during this semester then he will not pay for her tuition fee for the next semester.

Hanz: "If Kim studies hard on this semester then Carlo will pay for Kim's tuition fee next semester."

Betty: "If Carlo did not pay for Kim's tuition fee for this semester then Kim must've not studied hard last semester."

Whose reasoning is correct?

- **A.** Hanz
- **B.** Betty
- **C.** Hanz and Betty
- **D.** Neither Hanz nor Betty

Question: 19 of 40

When Emily told her friend Alex that she was considering skipping the morning run to focus on her work, Alex said: "You don't have to skip the run entirely. Regular exercise, such as running, helps improve focus and productivity."

Which of the following statement **strengthens** Alex's argument?

- **A.** Emily's work performance has been consistently poor.
- **B.** Alex is a professional athlete who manages both running and work responsibilities.
- **C.** Running is not effective for improving focus and productivity.
- **D.** Many successful professionals attribute their efficiency to regular exercise routines.

Question: 20 of 40

A language school offers an advanced level course for students who have completed the intermediate level or can hold conversations fluently. Students who excel in grammar and writing are invited to a special writing workshop. Among the students Jake, Emily, Liam, and Olivia participated.

Which conclusion is supported by the information?

- **A.** If Jake is invited to the writing workshop, he has completed the intermediate level.
- **B.** If Emily can hold conversations fluently, she is eligible for the advanced level course.
- **C.** If Liam completed the intermediate level, he is invited to the writing workshop.
- **D.** If Olivia is not invited to the writing workshop, she is in the intermediate level.

Question: 21 of 40

A kind of vitamin B3 is nicotinamide, often known as niacinamide. In addition to meat, fish, milk, eggs, green vegetables, and cereals, it may also be found in many other foods. For carbohydrates and fats to operate properly in the body and to keep cells healthy, niacinamide is necessary. A key component of skin whitening, niacinamide prevents the transfer of melanin pigment to skin cells. After being used for a while, niacinamide reduces hyperpigmentation and progressively lightens skin.

If the information is true, which of the following is **out of context** ?

- **A.** Niacinamide progressively lighten the skin and reduces hyperpigmentation.
- **B.** Niacinamide can be taken to genetically change any dark skin color.
- **C.** Niacinamide can be found on fish, milk, eggs, green vegetables, cereals and meat.
- **D.** Niacinamide is a kind of vitamin B3.

Question: 22 of 40

In a certain language, the word UNCLE is written as WPENG. How will you write the name AUNT?

- **A.** BVOU
- **B.** CWPV
- **C.** DXQW
- **D.** None of the above

Question: 23 of 40

Which one of the following words comes last in the dictionary?

- **A.** Achieve
- **B.** Acquit
- **C.** Accumulate
- **D.** Actuate

Question: 24 of 40

Which answer suits the given pattern?

ABDG, IJLO, QRTW, ...

1. EFHJ
2. GHJM
3. NPSW
4. KMOR

Question: 25 of 40

Emily ate a lot of sweets.
If Emily sleeps without brushing her teeth, she will likely to get toothache tomorrow.
If Emily got a toothache, she will go to the dentist.
If she goes to the dentist, she won't be able to participate in the lectures.

If the above information is true which of the following is impossible?

A. Emily went to the dentist but she did not brush her teeth before sleeping.

B. Emily didn't visit the dentist but she didn't attend the lectures too.

C. Emily brushed her teeth before sleep, but she went to the dentist.

D. Emily got a toothache but she participated in the lectures too.

Question: 26 of 40

There was a banner near the road displaying "**Speed thrills but kills.**"

Which one of the following statements supports the above slogan the best?

A. Driving at high speed is an action-packed experience

B. We should drive at a normal pace as high speed cause fatal accidents and reduce the chance of avoiding an accident, It may cost not only our life but even others

C. Watching high-speed vehicles are very thrilling, but waiting near a road is not safe as road accidents can be fatal.

D. Though driving at a high speed is a very thrilling experience, it's harmful and damages your vehicle a lot.

Question: 27 of 40

A paediatrician told: if your kid keeps a soft toy with her, don't forget to clean it daily. If your kid is suffering from asthma or other respiratory problems, then you should keep those stuffed toys away from your child. Otherwise, the teddy bear also can be harmful as a real bear.

Which of the following, if true, most strengthens the paediatrician's claim?

A. When a child loves teddy bears she might tend to think that real bears are not harmful too, which is not so good.

B. Soft toys quickly get dirty so it's essential to clean them daily otherwise the child will get sad and depressed.

C. Soft toys are home to dust mites and other allergens which are common triggers for allergy and other respiratory symptoms.

D. Soft toys and plush animals will help them reduce these feelings and stressful emotions.

Question: 28 of 40

Robbie started reading today's newspaper. There is an advertisement in first page that said :

"Buy pure and natural honey of company Golden Honey"

There are assuptions behind the above statement from advertisement.

Assumption I: No other company supplies pure and natural honey.
Assumption II: People read the advertisements.

Which assumption(s) is/are implicit ?

- **A.** Only assumption I is implicit.
- **B.** Only assumption II is implicit.
- **C.** Neither assumption I nor II is implicit.
- **D.** Both the assumptions are implicit.

Question: 29 of 40

All the sunflower in Ethan's garden is White.
All the Roses in Ethan's garden are Pink.
All the flowers in Ethan's garden are either White or Pink.

If the first two statements are true, the third statement is...

- **A.** True
- **B.** False
- **C.** Uncertain
- **D.** Can't decide

Question: 30 of 40

There are 53 students in a class. 20 students have more than 5 pens each and 15 students have 4 pens each and the remaining students have one pen each.

Melinda: "35 students have more than 2 pens each"
Harry: "The number of students not having more than 5 pens are 15"

Whose argument is correct?

- **A.** Melinda is correct
- **B.** Harry is correct
- **C.** Both Melinda and Harry is correct
- **D.** Neither Melinda nor Harry is correct

Question: 31 of 40

EVS, English and Maths books are kept in the manner explained.
Maths book is not the last book.
EVS book is above the maths book .
English book is not at second place.

Which book is the first book from the top?

- **A.** EVS
- **B.** Maths
- **C.** English
- **D.** Can't be determined.

Question: 32 of 40

Consider the following conditions regarding Samantha's marks.

If Samantha's marks are greater than 75, she will receive a gift.
If Samantha's marks are greater than 60, she can celebrate it.
If Samantha's marks are less than 65, she has to study well.
If Samantha's marks are less than 62, 2 penalty marks will be reduced.

Samantha got 62 marks for the exam.

Select the most suitable outcome.

1. Samantha can celebrate, but she will receive 2 penalty marks.
2. Samantha's new marks are 60.
3. Samantha will receive a gift, but she has to study well.
4. Samantha can celebrate, but she has to study well.

Question: 33 of 40

Handfish are a type of anglerfish famous for their **very unique** modified fins that resemble human hands, which they use to "walk" across the sea floor rather than swim.

Mandy: "If you see a fish with modified fins that resemble human hands it must be a Handfish"

Kara: "Also if you see a creature walk across the sea floor rather than swim, it must be a Handfish"

If the information given is correct, whose reasoning must be correct?

A. Mandy only
B. Kara only
C. Both Mandy and Kara
D. Neither Mandy and Kara

Question: 34 of 40

Find the first two terms of the following sequence.

A, B, 24, 26, 52, 54, 108, ...

1. 10, 20
2. 6, 12
3. 10, 12
4. 6, 8

Question: 35 of 40

Read the following statements carefully.

- If someone goes to the gym, then he becomes healthy.
- Healthy people are attractive.
- If someone is attractive, then he will have a lot of friends.

My friend Kevin goes to the gym.

Which answer is the **most correct** about Kevin?

1. Kevin is not attractive.

2. Kevin is healthy but he doesn't have a lot of friends.

3. Kevin is healthy and attractive, and he has a lot of friends.

4. Kevin's friends are attractive.

Question: 36 of 40

I have 10 friends. Five of them are vegetarians and two of them don't like ice cream.
Six of my friends have pet dogs and others have cats.

Which of the following is NOT correct about my friends?

1. Some of my friends have pet dogs.

2. If someone is a friend of mine, he likes ice cream.

3. I have some friends who like ice creams.

4. All of my friends have dogs or cats.

Question: 37 of 40

Consider the following statements.

Tasty fruits are fragrant and contain vitamin C.
James found a fruit from a forest and it's fragrant.

Select the correct conclusion from the above statements.

1. If it contains vitamin C, it is tasty.

2. The fruit is not tasty.

3. If it does not contain vitamin C, it is not tasty.

4. It is tasty because it is fragrant.

Question: 38 of 40

If 2-3-5-8-1-6 is GROUND and 8-9-0-0 is BALL, what is the meaning of 8-3-4-2-9-6-7?

A. BRIGADE

B. BRING

C. BRIGADIER

D. BRIGAND

Question: 39 of 40

Antioxidants and minerals like calcium and potassium are abundant in radishes. These nutrients work as a team to lower high blood pressure and lower your risk of developing heart disease. The radish is a good source of naturally occurring nitrates that enhance blood circulation.

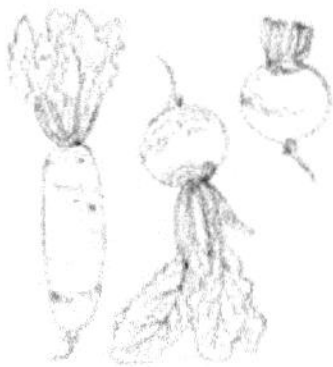

If the information is true, which of the following must be false?

- **A.** Radish enhances blood circulation.
- **B.** Nutrients in radish lowers high blood pressure.
- **C.** Radish are abundant in calcium and potassium.
- **D.** The antioxidants and minerals in radish increases the risk of developing a heart disease.

Question: 40 of 40

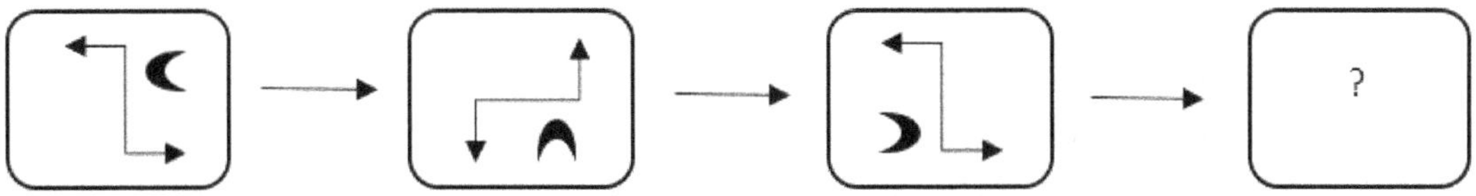

A.

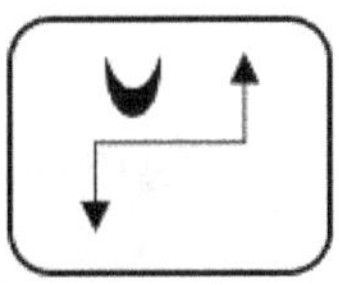

B.

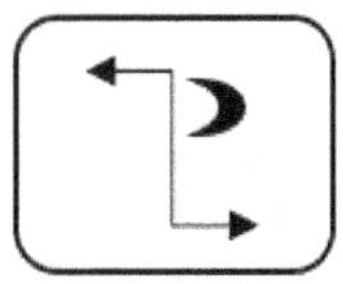

C.

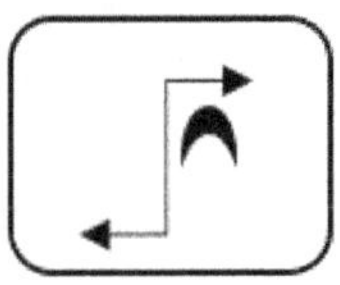

D.

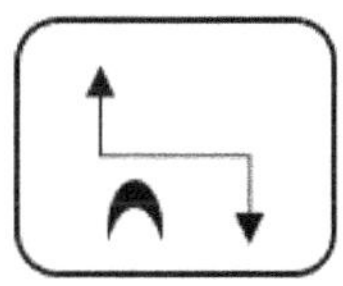

Q: 1 Who is credited with inventing the telephone?

A. Thomas Edison

✓ Correct Ans **B.** Alexander Graham Bell

C. John Logie Baird

D. Samuel Morse

Explanation

According to Extract A, Alexander Graham Bell is credited with inventing the telephone.

Q: 2 How has the telephone impacted communication today?

A. It has led to a decrease in face-to-face communication.

B. It has made communication less convenient.

C. It has led to an increase in letter writing.

✓ Correct Ans **D.** It has changed how people connect and interact with each other.

Explanation

According to Extract A, the telephone has changed how people connect and interact with each other through features like texting, voice calls, and video conferencing.

Q: 3 According to the United Nations Conference on Trade and Development (UNCTAD), which country in the Gulf region has the highest proportion of mobile phone users?

✓ Correct Ans **A.** Saudi Arabia

B. Oman

C. Kuwait

D. UAE

Explanation

According to Extract B, Saudi Arabia has the greatest percentage of mobile users among Gulf region countries, according to a study by the United Nations Conference on Trade and Development (UNCTAD).

Q: 4 What did the cohort study in Denmark find regarding the relationship between cell phone use and brain tumors?

A. There is a significant association between the two.

✓ Correct Ans **B.** There is no association between the two.

C. There is a weak association between the two.

D. There is a potential association between the two.

Explanation

According to Extract B, there is no evidence to suggest that mobile phone use increases the risk of developing brain tumors.

Q: 5 According to the FDA, how can individuals reduce their exposure to radiofrequency energy when using a cell phone?

A. Use a hands-free device

B. Use the phone for shorter conversations or when a landline is available

✓ Correct Ans **C.** Both A and B

D. Neither A nor B

Explanation

The FDA has recommended the following measures to minimize radio frequency energy exposure: Cell phones should only be utilized during brief conversations or in situations where a landline phone is unavailable. By employing a hands-free device, the user effectively increases the distance between the phone and their head.

Q: 6 In which year do the late chemist James Lovelock—then a researcher at NASA's Jet Propulsion Laboratory—devise a feasible strategy for detecting life on other worlds?

✓ Correct Ans **A.** In 1965

B. In 1956

C. In 1960

D. In 1985

Explanation

According to third paragraph, in 1965 the late chemist James Lovelock—then a researcher at NASA's Jet Propulsion Laboratory—devised a feasible strategy for detecting life on other worlds. This supports option A as the answer to the question.

Q: 7 Which of the following was Peter Ward's view on life on Mars?

A. Certain chemical compounds in a planet's atmosphere act as so-called biosignatures indicating life's global presence

✓ Correct Ans **B.** Life is more self-destructive than self-regulating and eventually wipes itself out

C. The flood of atmospheric oxygen proved crucial for our planet's biological diversification

D. None of the above

Explanation

According to fourth paragraph, Peter Ward of the University of Washington put forward a less optimistic view. At planetary scales, Ward argued, life is more self-destructive than self-regulating and eventually wipes itself out, This supports option B as the answer to the question.

Q: 8 Through the article, we can conclude that researchers ...

✓ Correct Ans **A.** have yet to find any direct proof that life of any kind ever graced the Martian surface.

B. have successfully found proof of life on Mars.

C. have found that Mars is habitable by 2099.

D. have successfully found a new discovery that is secret to human.

Explanation

Accordng to first paragraph of the text, researchers have yet to find any direct proof that life of any kind ever graced the Martian surface. This supports option A as a correct statement.

Q: 9 Which of the following describes Mars?

A. Same as earth

B. Humans can stay on Mars in 2099

C. Researchers found signs of life on Mars.

✓ Correct Ans **D.** Wetter, warmer, and more habitable than the freeze-dried desert world of today

Explanation

According to first paragraph of the text, Mars was wetter, warmer, and more habitable than the freeze-dried desert world of today, researchers have yet to find any direct proof that life of any kind ever graced the Martian surface. This supports option D as the best answer to the question.

Q: 10 What is the primary theme of the poem "Barter" by Sara Teasdale?

A. The importance of material wealth

✓ Correct Ans **B.** The value of life's loveliness

C. The pursuit of happiness

D. The fleeting nature of beauty

Explanation

The primary theme of the poem is the value of life's loveliness, as it encourages the reader to appreciate and cherish beautiful moments in life.

Q: 11 What poetic device is used in the line "Music like a curve of gold"?

✓ Correct Ans **A.** Simile

B. Metaphor

C. Alliteration

D. Onomatopoeia

Explanation

The line "Music like a curve of gold" uses a simile to compare music to a curve of gold, emphasizing its beauty and preciousness.

Q: 12 In the line "Spend all you have for loveliness," what does the speaker advise the reader to do?

A. Save money for the future

B. Prioritize material possessions

✓ Correct Ans **C.** Cherish and invest in life's beauty

D. Live a frugal life

Explanation

The speaker advises the reader to prioritize and invest in life's loveliness, rather than material possessions.

Q: 13 What is the effect of the phrase "holy thoughts that star the night" in the poem?

A. It suggests a religious theme.

✓ Correct Ans **B.** It conveys a sense of spiritual or transcendent beauty.

C. It emphasizes the darkness of the night.

D. It symbolizes the end of the day.

Explanation

The phrase "holy thoughts that star the night" conveys a sense of spiritual or transcendent beauty, suggesting that such thoughts brighten the darkness of night.

Q: 14 In the context of the poem, what does the phrase "never count the cost" mean?

A. Keep track of expenses

B. Ignore the value of beauty

C. Prioritize financial gain

✓ Correct Ans **D.** Do not hesitate to invest in loveliness

Explanation

In the context of the poem, "never count the cost" means not hesitating to invest in loveliness and not focusing on material expenses when it comes to experiencing beauty.

Q: 15 Which of the extracts refers to the time when leaves start to change color ?

A. Extract A

B. Extract B

C. Extract C

✓ Correct Ans **D.** Extract D

Explanation

The second line of extract D proves that fact.

Q: 16 Which of the extracts refers to the cozy nights by the fire?

A. Extract A

✓ Correct Ans **B.** Extract B

C. Extract C

D. Extract D

Explanation

The paragraph states that It is a time when the world transforms into a winter wonderland, with snow-covered landscapes and cozy nights by the fire.

Q: 17 Which of the extracts states about the season when the people go on vacations?

✓ Correct Ans **A.** Extract A

B. Extract B

C. Extract C

D. Extract D

Explanation

Extract A describes that many people go on vacations during this time, exploring new places and creating lasting memories

Q: 18 Which of the extracts refers to the season which reminds us of the cyclical nature of life?

A. Extract A

B. Extract B

C. Extract C

✓ Correct Ans **D.** Extract D

Explanation

Extract D states that Autumn is a season that reminds us of the beauty in change and the cyclical nature of life.

Q: 19 Which one fits in (1) ?

A. Sentence A

B. Sentence B

✓ Correct Ans **C.** Sentence G

D. Sentence H

Explanation

The essay is about tornadoes and the correct answer should be he introductory line to that.

Q: 20 Which one fits in (2) ?

✓ Correct Ans **A.** Sentence A

B. Sentence B

C. Sentence C

D. Sentence D

Explanation

The prior sentence states about air collision and the correct answer also states about a collision.

Q: 21 Which one fits in (3) ?

A. Sentence A

B. Sentence D

✓ Correct Ans **C.** Sentence B

D. Sentence E

Explanation

The paragraph is about the types of tornadoes.

Q: 22 Which one fits in (4) ?

A. Sentence A

B. Sentence B

C. Sentence H

✓ Correct Ans **D.** Sentence E

Explanation

The prior sentence mentions about the destructive nature of tornadoes. Therefore, the correct answer carries an example for that.

Q: 23 Which one fits in (5) ?

A. Sentence A

B. Sentence B

✓ Correct Ans **C.** Sentence C

D. Sentence D

Explanation

The correct answer should be tally with the prior sentence of precautions.

Q: 24 Which one fits in (6) ?

A. Sentence A

B. Sentence B

C. Sentence C

✓ Correct Ans **D.** Sentence D

Explanation

The correct answer should support the prior sentence about the areas where tornadoes commonly happens.

Q: 25 Which one fits in (7) ?

A. Sentence A

B. Sentence G

✓ Correct Ans **C.** Sentence H

D. Sentence C

Explanation

The correct answer should be related to research and technology

Q: 26 What is the predominant tone of the poem "The House on the Hill"?

A. Joyful

✓ Correct Ans B. Melancholic

C. Excited

D. Optimistic

Explanation

The predominant tone of the poem is melancholic, as it conveys a sense of loss and desolation.

Q: 27 What is the significance of the repeated line, "There is nothing more to say," in the poem?

A. It emphasizes the abundance of words spoken.

✓ Correct Ans B. It highlights the finality and emptiness of the situation.

C. It suggests that there are unresolved issues.

D. It indicates the presence of ongoing conversations.

Explanation

The repeated line, "There is nothing more to say," underscores the finality and emptiness of the situation, emphasizing that there are no words left to express.

Q: 28 What deeper meaning the House on the Hill symbolize in the poem?

A. A place of happiness and celebration

B. A symbol of eternal life

✓ Correct Ans C. The passage of time and the inevitability of change

D. A representation of endless possibilities

Explanation

The House on the Hill likely symbolizes the passage of time and the inevitability of change, as it is described as in a state of ruin and decay.

Q: 29 How does the poet convey a sense of abandonment in the poem?

A. Through vivid descriptions of the house's interior

B. By emphasizing the presence of people inside the house

✓ Correct Ans C. Through the use of repetition and the line "They are all gone away"

D. By portraying the house as a welcoming and lively place

Explanation

The poet conveys a sense of abandonment through the use of repetition and the line "They are all gone away," which suggests that the house is empty.

Q: 30 What does the phrase "our poor fancy-play" in the poem imply?

A. The imagination's ability to create stories

B. The importance of creativity in life

✓ Correct Ans **C.** The futility of artistic endeavors

D. The joy of childhood games

Explanation

The phrase "our poor fancy-play" implies the futility or insignificance of artistic endeavors, suggesting that their creative efforts were in vain.

Answers And Solutions
Learnikx Education

Q: 1 The difference between the present ages of Arun and Deepak is 14 years. Seven years ago the ratio of their ages was 5:7
.
What is Deepak's present age?

A. 49 years

B. 42 years

C. 63 years

✓ Correct Ans **D.** None of these

Explanation

Seven years ago, Arun was 5n and Deepak was 7n.
Now, Arun = 5n + 7, Deepak = 7n + 7
Difference = 2n = 14; n = 7

Deepak's present age = 7x7 + 7 = 56

Q: 2 Robert walks slower than Cooper and Cooper walks as fast as Collin and Alan walk faster than Collin.
Who walks the fastest?

A. Cooper

✓ Correct Ans **B.** Alan

C. Robert

D. Both Cooper and Collin

Explanation

Robert < Cooper
Cooper = Collin
Alan > Collin

So the final order is :
Robert < Cooper/Collin < Alan

Q: 3 There are 4 red buttons, 6 white buttons and 3 black buttons in an invisible cord.
What is the minimum number of buttons that must be drawn one at a time to draw **at least one black button everytime** without looking at the button?

A. 6

B. 8

✓ Correct Ans **C.** 11

D. 15

Explanation

Add the number of red and white buttons and add 1 more to it.
4 +6 +1 = 11
The correct answer is: C. 11

Q: 4 □ represents one number.

△ represents another different number.

If :
7 X □ = 112
△ + □ = 96

What is △?

A. 10

B. 20

C. 42

✓ Correct Ans **D.** 80

Explanation

Using the fact that 7 X □ = 112, we can work out that □ = 16, because 112÷7=16.
△ + □ = 96 ; so △ = 96 - 16 ; △ = 80.
So the correct answer is D. 80.

Q: 5 How many minimum straight lines should be drown inside the figure to make **six small same-size rectangles** ?

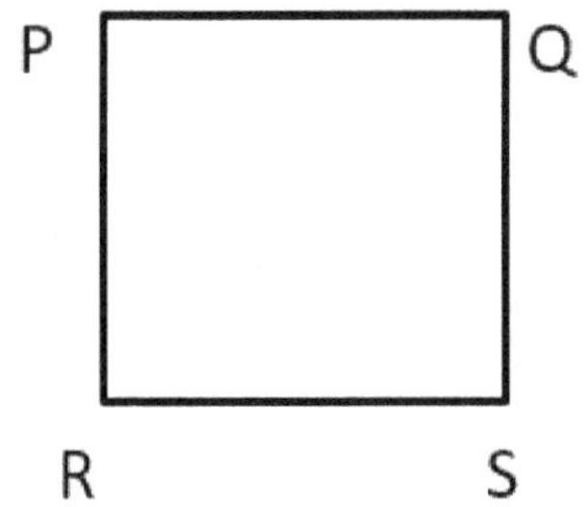

A. 6

✓ Correct Ans **B.** 3

C. 2

D. 4

Explanation

Let's see how many straight lines can be drawn.

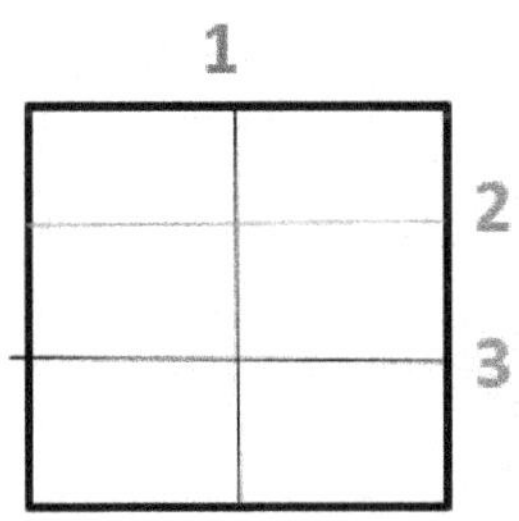

The correct answer is: B. 3

Q: 6 In a hall, there are 85 rows of chairs and each row has 24 chairs. Two of those rows of chairs are brought into an empty classroom and to be re-arranged into rows of 8. How many rows are there in the classroom now?

A. 48

B. 8

C. 9

✓ Correct Ans **D.** 6

Explanation

24 x 2 = 48
48 ÷ 8 = 6
Note : There is too much information in the question.

Q: 7 Khairul and Jason had a total of 1232 cards. After Khairul gave 342 cards to Jason, Khairul still had 298 more cards than Jason.
How many cards did Jason have at first?

A. 250

✓ Correct Ans **B.** 125

C. 225

D. 230

Explanation

At start : Jason = n, Khairul = 1232 - n
At end : Jason = n + 342, Khairul = 1232 - n - 342 = 890 - n

n + 342 + 298 = 890 - n
n + 640 = 890 - n
n x 2 = 890 - 640 = 250; n = 125

Q: 8 There was an equal number of ping pong balls in Box A and Box B at first.
After 47 ping pong balls were removed from Box B and 47 ping pong balls were added in Box A, there were 236 ping pong balls in the two boxes altogether.
How many ping pong balls were in Box B in the end?

A. 156

B. 165

✓ Correct Ans **C.** 71

D. 175

Explanation

At start : Box A = n, Box B = n
At the end: Box A: n + 47, Box B: n-47, 2n = 236, n = 118

At the end : Box B = n - 47 = 118 - 47 = 71

Q: 9 There were 20 400 registered voters for a certain election. 2% of them did not come for voting.
How many voters did come for voting?

A. 16 992

B. 20 000

C. 15 598

✓ Correct Ans **D.** 19 992

Explanation

Let's find the percentage of voters.
Voters came for voting =100% -2% = 98%
If there are 20400 registered voters, let's find out number of came to vote.
98% multiply by registered voters

$$20\,400 \times \frac{98}{100} = 19\,992$$

So correct answer: D. 19 992

Q: 10 Jenny earned $3 for every box of food that she delivered. She received an additional $20 for every 10 boxes of food delivered. How many boxes of food did she deliver if she earned a total of $218?

A. 22

B. 120

C. 9

✓ Correct Ans **D.** 46

Explanation

$3 X 10= $30
$30 + $20 = $50
So one group is $50.

How many groups ? $218 ÷ $50 = 4 r $18
$18 ÷ $3 = 6

No. of boxes = (4 X 10) + 6 = 46

Q: 11

A baker bought some eggs. She used $\frac{1}{3}$ of the eggs to bake tarts, 92 eggs to bake muffins and $\frac{2}{3}$ of the remaining eggs to bake cupcakes. After that, she still had 36 eggs. How many eggs did the baker buy?

A. 525

B. 309

✓ Correct Ans **C.** 300

D. 209

Explanation

Strategy : working backwards
cupcakes + left = 36 X 3 = 108
cupcakes + left + muffins = 108 + 92 = 200
$^{2}/_{3}$ of total = 200
Total = 300

Q: 12

During a sale, cooking pots and frying pans were sold at the prices shown in the table below.

Cooking Pot	Frying Pan

Buy 1 at $810.90	Buy 1 at $640
Buy 2 or more at $800 each	Buy 2 or more at $630.50 each

Naomi bought 1 cooking pot and 4 frying pans during the sale. What was the smallest amount of money she paid?

A. $4354.60

✓ Correct Ans **B.** $3332.90

C. $3342.80

D. $5445.10

Explanation

Amount spent → $810.90 + ($630.50 X 4) = $3332.90

Q: 13 Kelly woke up 20 minutes earlier than usual. As she woke up early, the time she took to get ready is delayed by 6 minutes. Usually, she arrives at school exactly at 7.05 am.
Mcki woke up 10 minutes later than usual. Usually, he arrives at school 18 minutes earlier than Kelly.
Calculate the time difference between Kelly and Mcki arriving at school that day.

✓ Correct Ans **A.** 6 minutes

B. 5 minutes

C. 10 minutes

D. 7 minutes

Explanation

Mcki = 7.05 am – 18 minutes = 6.47am
=> 10 minutes late = 6.47am + 10 minutes = 6.57am

Kelly = the time difference of waking up early but delayed 6 minutes to get ready
=> (20 mins – 6 mins = 14)= 7.05 am – 14 mins = 6.51am

The time difference = 6.57 am – 6.51 am = 6 minutes

Q: 14 Number 85 can be split in two perfect squares into two different ways. That is: 85 = 81 + 4 or 85 = 49 + 36.
Select another number in the given choices that is less than 100 and can be split into two different squares.

A. 36

B. 40

✓ Correct Ans **C.** 25

D. 72

Explanation

25 is the number that can be split into different squares. 25 = 16 + 9. 16 is formed by 4^2 and 9 results from 3^2.

Q: 15 The mean temperature for 7 days is 24.60˚C. For the first six days, the mean temperature was 21.60˚ C. What was the temperature on the 7th day?

✓ Correct Ans **A.** 42.60˚C.

B. 36.60˚C.

C. 24.60˚C.

D. 21.60˚C.

Explanation

The mean temperature for 7 days is 24.60˚C; total = 172.2˚C.
The mean temperature for 6 days is 21.60˚C ; total = 129.60˚C.

Total temp. for 7 days - total temp. for 6 days = 172.20˚C - 129.60˚ C = 42.60˚C.

Q: 16 May had three times as much money as June. May spent $1700. As a result, she had $300 less than June. How much money did May have left?

A. $449

B. $440

C. $350

✓ Correct Ans **D.** $400

Explanation

At start : May = nx3, June = n
At end : May = nx3 - 1700, June = n

nx3 - 1700 = n - 300
nx2 = 1400 ; n = 700
May is left with = nx3 - 1700 = 2100 - 1700 = 400

Q: 17 Miss Sharma baked some pies to be packed into boxes.
If she packed 5 pies in each box, she would have 4 pies left. If she packed 9 pies in each box, she would have 8 pies left.
Given that Miss Sharma baked fewer than 60 pies, how many pies did Miss Sharma bake?

✓ Correct Ans **A.** 44

B. 48

C. 54

D. 24

Explanation

Make a numbered series.

5 times table + 4 : 9, 14, 19, 24, 29, 34, 39, **44**, 49, 54, 59.
9 times table + 8 : 17, 26, 35, **44**, 53.

44 is common.

Q: 18 Karen went to the post office to send some letters to her friends in Japan. The postage rates to Japan is shown In the table below.

Mass of letter	
First 30 g	$1.50
Every additional 10 g or part thereof	40 ¢

Karen wanted to send a letter with a mass of 55 g to Japan. She gave the cashier a $5 note. How much change did she get?

A. $4.00

✓ Correct Ans **B.** $2.30

C. $3.50

D. $6.50

Explanation

She had to pay = $1.50 + ($0.40 x 3) = $2.70
She received = $5 - $2.70 = $2.30

Q: 19 A small square of side 6 cm is cut out from a bigger square of area 144 cm^2. Find the ratio of the shaded area to that of the unshaded area in the simplest form.

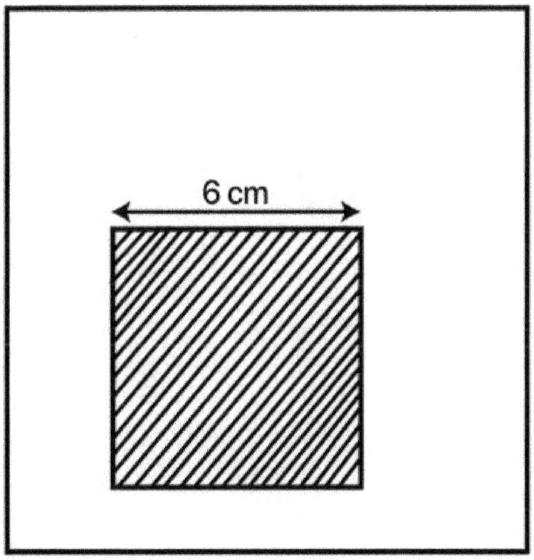

A. 1:5

B. 3:1

✓ Correct Ans **C.** 1:3

D. 2:5

Explanation

Shaded = 6 x 6 = 36cm^2
36 : 144 - 36 = 36 : 108 = 1 : 3

Q: 20

There are some marbles in a container. $\frac{1}{3}$ of the marbles are blue. $\frac{1}{4}$ of the remaining marbles are yellow and the rest are red.
What fraction of the marbles in the container are yellow?

A. 6

✓ Correct Ans **B.** $^1/_6$

C. $^1/_3$

D. $^7/_{16}$

Explanation

$^2/_3 \times {}^1/_4 = {}^2/_{12} = {}^1/_6$

Q: 21

Terry read $\frac{1}{5}$ of the number of pages in a book in the first hour, $\frac{3}{8}$ of the remaining pages in the second hour. After that, he still had 45 more pages to read.
How many pages are there in the book?

A. 19

✓ Correct Ans **B.** 90

C. 15

D. 99

Explanation

Remaining : $^4/_5 \times {}^5/_8 = 1/2$

$^1/_2$ is 45 pages. So whole is 90 pages.

Q: 22 The pie chart shows the results of a survey conducted to identify the favorite game of some students.

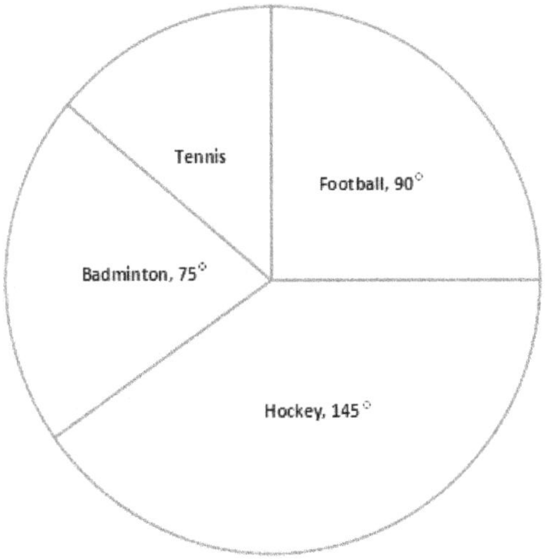

How many students like Tennis if the total number of students is 1,260?

✓ Correct Ans **A.** 175

B. 90

C. 155

D. 65

Explanation

Tennis = 360 – 75 – 145 – 90 = 50
$^{50}/_{360}$ x 1,260 = 175 students

Q: 23 The sides of a triangle are in the ratio : $\frac{1}{2}:\frac{1}{3}:\frac{1}{4}$

If the perimeter of the triangle is 52 cm, what is the length of the shortest side?

A. 9 cm

B. 10 cm

C. 11 cm

✓ Correct Ans **D.** 12 cm

Explanation

Since the sides of the triangle are in the ratio $^1/_2$:$^1/_3$:$^1/_4$ multiply the three by 12.
$^1/_2$ x 12 : $^1/_3$ x 12 : $^1/_4$ x 12 = 6:4:3
Let the sides be 6x cm, 4x cm, and 3x cm. The perimeter is 52 cm
Therefore, 6x + 4x + 3x = 52 => 13x = 52 => x = 4
Therefore, the length of the shortest side is 3x cm = 12 cm

Q: 24 Tomek, Romek, Andrew, and Michael said the following about a certain number:

***Tomek**: "This number is equal to 9."*
***Romek**: "This number is prime."*
***Andrew**: "This number is even."*
***Michael**: "This number is equal to 15."*

Only one statement given by either Romek or Tomek is true,
Only one statement given by either Andrew or Michael is true.
What number is it?

✓ Correct Ans **A.** 2

B. 3

C. 9

D. 15

Explanation

If the number is 2, then:
Tomek's statement was false, but Romek's statement was true (2 is prime)
Andrew's statement was true, but Michael's statement was false

If the number is 3, then:
Tomek's statement was false, but Romek's statement was true (3 is prime)
Andrew's statement was false, and Michael's statement was also false

If the number is 9, then:
Tomek's statement was true, but Romek's statement was false (9 is not prime)
Andrew's statement was false, and Michael's statement was also false

If the number is 15, then:
Tomek's statement was false, and Romek's statement was also false (15 is not prime)
Andrew's statement was false, but Michael's statement was true

2 is the only answer satisfying the condition that only one statement given either by Romek or Tomek is true, as well as only one statement given by either Andrew or Michael is true.

Q: 25 The total cost of a pen and 2 similar highlighters is $6.50, The total cost of 2 such pens and 1 such highlighter is $7. What is the cost of 5 such highlighters?

A. $13

✓ Correct Ans **B.** $10

C. $9

D. $6

Explanation

pen + 2 highlighter = $6.50 .. (1)
2 pen + 1 highlighter = $7 .. (2)

2 pen + 4 highlighter = $6.50 x 2 = $13 .. (3)
From (2) and (3) :
3h = $13 - $7 = $6
1h = $6 ÷ 3 = $2
5h = 5 x $2 = $10

Q: 26

Aaron has $\frac{3}{5}$ as much savings as Bob. If Bob gives $40 to Aaron, they will have the same amount of money. How much money does Bob have?

A. $350

B. $225

✓ Correct Ans **C.** $200

D. $180

Explanation

At start : Aaron = 3n, Bob = 5n
At end : Aaron = 3n + 40 , Bob = 5n - 40
3n + 40 = 5n - 40; 2n = 80; n = 40
Bob has = 5n = 40 x 5 = 200

Q: 27

The bar graph below shows the number of old textbooks donated by five Primary 5 classes to a children's home.

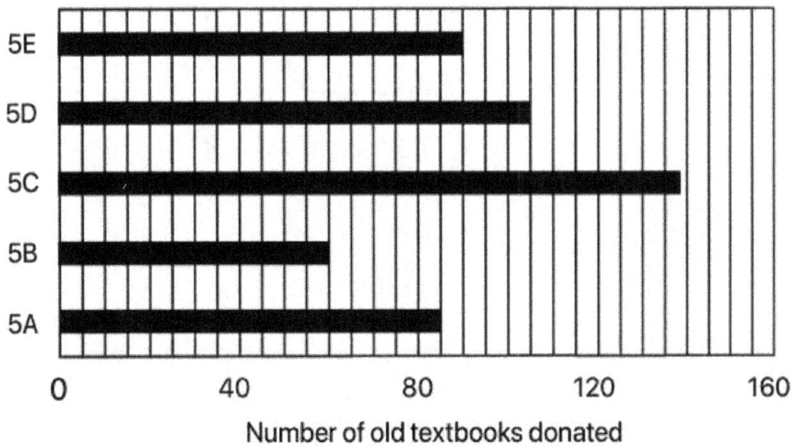

What fraction of the old textbooks was donated by class 5E? Give your answer in the simplest form.

A. $^5/_{16}$

B. $^3/_7$

✓ Correct Ans **C.** $^3/_{16}$

D. $^3/_8$

Explanation

90 + 105 + 140 + 60 + 85 = 480

90 ÷ 480 = $\frac{3}{16}$

Ans: $\frac{3}{16}$

Q: 28 In the figure, ABCD is a rectangle where AC and BD are straight fines. Given that $\angle AEB = 130°$ and AE = EC = DE = EB, find $\angle ADB$.

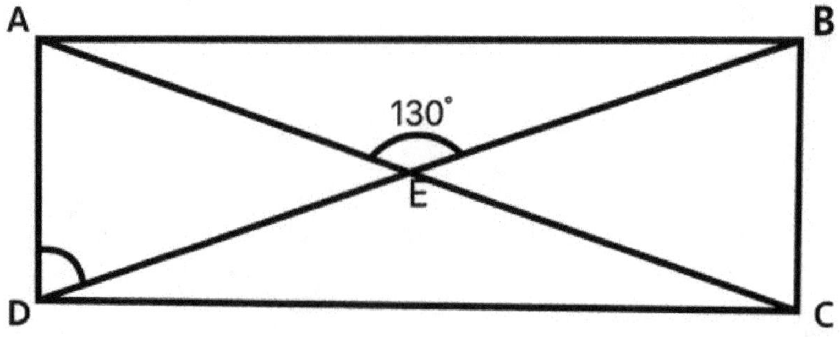

A. 75°

✓ Correct Ans **B.** 65°

C. 130°

D. 50°

Explanation

360 - (130° x 2) = 100°
100° ÷ 2 = 50°

180°- 50° = 130°
130° ÷ 2 = 65°

Q: 29 @, #, $ represent 3 numbers.

#-$=7
@/6+$=#-2
$ ×11=121

What is the value of @?

A. 18

B. 24

✓ Correct Ans **C.** 30

D. 42

Explanation

By using 3rd equation :
$ ×11= 121
$=121/11=11
Use $=11 in 1st equation :
#-$=7
#=7+11 =18
From 2nd equation :
@/6+$=#-2
@/6+11=18-2
@/6=16-11
@=5×6 =30

Q: 30 A large rectangle has side lengths of 11 cm and 21 cm. Three identical rectangles of length 5 cm and width 2 cm are removed. It leaves the below figure.

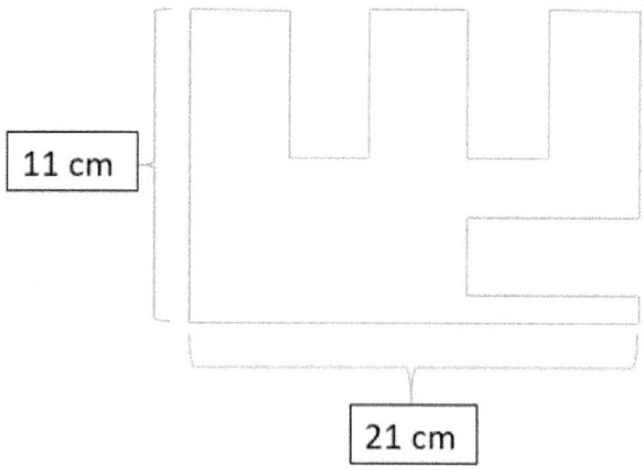

What is the perimeter of the given figure?

A. 64cm

B. 76cm

C. 88cm

✓ Correct Ans **D.** 94cm

Explanation

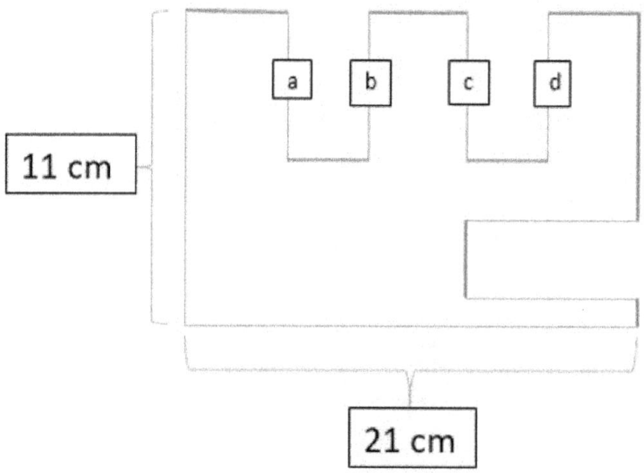

The highlighted horizontal lines must be added up to 21 cm and the highlighted vertical lines must be added up to 11 cm.

So the perimeter of the figure is 11+21+11+21+a+b+c+d+e+f

Since the length of the removed small rectangle is equal to 5 cm, a=b=c=d=e=f=5

Perimeter= 11+21+11+21+ (5 ×6) =94cm

Q: 31 Some money was given to three daughters A, B and C by their mother to buy household things. The total amount of money given was \$4350.
A spent $^{6}/_{15}$ and B spent $^{17}/_{30}$ of the total amount and C spent the remaining.

Which of the following statement(s) is/are correct?

1. The highest amount was spent by A.
2. Amount spent by C was $^{1}/_{30}{}^{th}$ of the total amount.
3. Amount spent by B was more than 2 quarters of the total amount.

A. Statement 1 only

B. Statement 2 only

C. Statements 1 and 2 only

✓ Correct Ans **D.** Statements 2 and 3 only

Explanation

Amount spent by A=\$4350 ×$^{6}/_{15}$ =\$1740
Amount spent by B=\$4350 × $^{17}/_{30}$ =\$2465
Amount spent by C=\$4350- (1740 + 2465) =\$4350- \$4205 =\$145

1. Highest amount was spent by B and it was \$2465. Statement 1 is incorrect.
2. $^{1}/_{30}$ of the total amount is (\$4350 × $^{1}/_{30}$)=\$ 145 and it is equal to the amount spent by C. Statement 2 is correct.
3. Two quarters of the total amount is equal to ½ of the total amount and it is \$4350 ×$^{1}/_{2}$=\$2175 Amount spent by B is \$2465 and it is greater than \$2175. So the amount spent by B is greater than 2 quarters of the total amount. Statement 3 is correct.

Q: 32

50	21	12	A
62	52	81	10
28	39	95	43
80	11	03	111

Identify the pattern and find A.

A. 64

B. 88

C. 111

✓ Correct Ans **D.** 122

Explanation

Consider all the numbers in each horizontal line. Add them. They all get the sum of 205 when adding separately.
62+52+81+10=205
28+39+95+43=205
80+11+3+111=205
So the 1st raw also must add up to 205.
So the answer is,
A= 205-(50+21+12) = 122
No pattern found when adding numbers in vertical lines.
50+62+28+80= 220
21+52+39+11=123
12+81+95+3= 191
So the answer is D.

Q: 33 Arthur and Merlin did the game of throwing an unbiased cube shaped dice.
Arthur threw the dice first and got number 3. Then Merlin threw the dice once.

Which of these statement(s) is/are correct?

A. *The probability of Arthur getting 3 is* $^1/_6$.

B. *The probability of Merlin getting a number greater than Arthur is* $^1/_2$.

C. *The probability that Merlin and Arthur's scores having a difference less than 3 is* $^5/_6$.

A. Only A

B. Only A and B

C. Only B and C

✓ Correct Ans **D.** All A, B and C

Explanation

A. There are 6 equal sides in a cube shaped dice. As those sides are numbered from 1 to 6, probability of Arthur getting 3 is $^1/_6$. Statement A is correct.

B. Numbers greater than Arthur's are : 4, 5 and 6. Accordingly there are 3 outcomes for that. So the probability of Merlin getting a number greater than Arthur's is $^3/_6 = ^1/_2$. Statement B is correct.

C. Arthur got a 3.
Merlin gets a 1: difference=3-1=2
Merlin gets a 2: difference=3-2=1
Merlin gets a 3: difference=3-3=0
Merlin gets a 4: difference=4-3=1
Merlin gets a 5: difference=5-3=2
Merlin gets a 6: difference=6-3=3
Except when getting number 6, the difference is less than 3.
So probability is $^5/_6$
Statement C is correct.

Q: 34 The following pie chart shows unsold food (in Kgs) in the canteen at close of business.

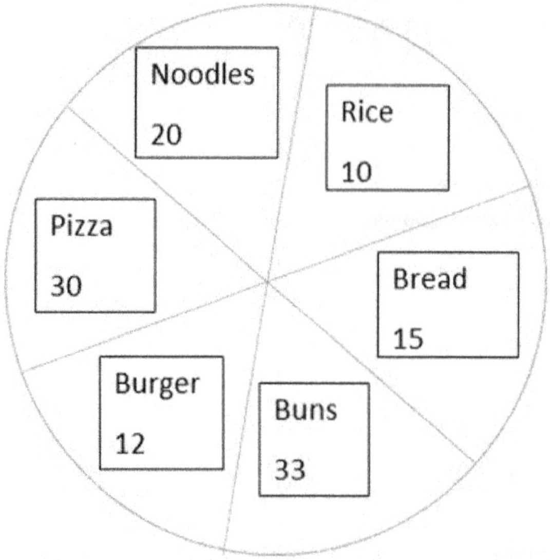

What should be the angle at the center for burger and pizza respectively?

A. 90°, 36°

B. 45°, 60°

✓ Correct Ans **C.** 36°, 90°

D. 60°, 90°

Explanation

Total of remaining foods= 20+10+15+33+12+30 =120

Angle for burger= $^{12}/_{120}$ × 360° = 36°

Angle for pizza= $^{30}/_{120}$ × 360 ° = 90°

Q: 35 The below bar graph shows favourite books of students in a high school.

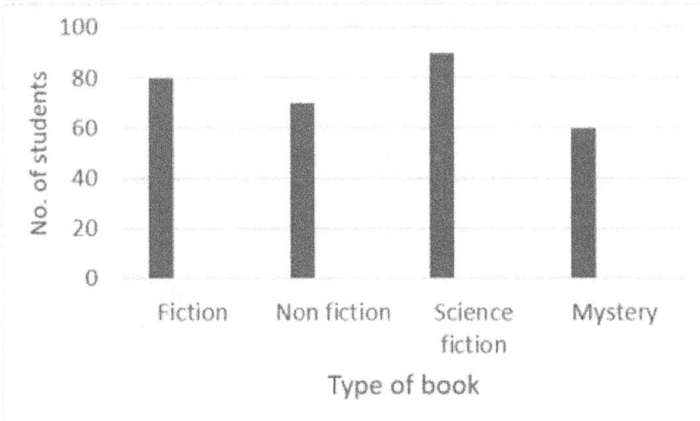

If above information were shown in a pie chart, what would be the difference of the angles at the center of fiction and mystery books?

A. 45°

B. 38°

✓ Correct Ans **C.** 24°

D. 16°

Explanation

Total no. of students= 80+70+90+60 = 300
Angle for fictions= 80/300 ×360 ° = 96°
Angle for mystery books= 60/300 ×360 ° =72°
Angle difference= 96 °-72 ° = 24°

Q: 1 In a railway station, there is a bulb indicator system for train arrivals and departures. Whenever the train leaves the nearest station, the yellow bulb turns on to indicate that the train will arrive soon. And when the train is going to leave the station within 10 minutes, the red light starts to flash on and off.

Richard: "The red bulb is on continuously, which means that the train has already left. We have to wait for another train"

Samson: " The yellow bulb turned on, which means another train will arrive soon. Let's wait."

If the given information is correct, whose reasoning must be correct?

A. Richard only

✓ Correct Ans **B.** Samson only

C. Both Richard and Samson.

D. Neither Richerd nor Samson.

Explanation

No information is given about the meaning of continuous red light, and nor does it follow from the meaning of a flashing light, so Richards is incorrect.
Whenever the train leaves the nearest station, the yellow bulb turns on to indicate that the train will arrive soon. So Samson's reasoning is correct.

Q: 2 Parasitic plants are plants that depend on other plants for nutrients. All parasitic plants are divided into two types. Partial parasites are those that depend on the host plant only for water and minerals while total parasites depend on the host plant for water, minerals and food too. Cuscuta which is commonly known as Dodder is a parasitic plant. Since they lack chlorophyll, they cannot use photosynthesis to produce own food.

If the given information is correct, which one of the following sentences is **not correct**?

A. Cuscuta must be a total parasitic plant.

✓ Correct Ans **B.** Cuscuta plants depend on the host plant only for water and minerals.

C. Cuscuta plant depends on the host plant for all water, minerals and food.

D. Partial parasites don't absorb food from the host plant.

Explanation

Cuscuta is a parasitic plant. Since it can't produce its own food it must depend on the host plant for food. So it must be a total parasite. Total parasites depend on the host plant for all water and minerals and foods. Therefore B is not correct.

Q: 3 **Whoever made the choreography of the first-place winning dance must have unique ideas and a keen passion.**

If this is true, which one of these statements must also be true?

A. If Liam has unique ideas and a keen passion, he must be the choreographer of the first-place-winning dance.

B. Liam is a choreographer, and if he has unique ideas and a keen passion then he must be the person who made the choreography of the first-place winning dance

C. If Liam is not the person who made the choreography of the first-place-winning dance, he cannot have unique ideas.

✓ Correct Ans **D.** If Liam doesn't have unique ideas, he cannot be the choreographer of the first winning dance.

Explanation

If the person made the choreography of the first-place winning dance must have unique ideas and a keen passion, then anyone who does not have either of those cannot be the person who made the choreography of the first-place winning dance.

Q: 4 In a team game, they have to draw lots first. There are two teams. The team who got blue colour lot have to swim through the lake first and then have to follow the route and find the hidden map. The team members who got the red lot first had to go through the mud pit and then find the hidden map. Both hidden maps show the path to the same destination. The team that reaches the destination first will win the game.

Team A got the red lot, but they lost the game.

Which of the following **cannot** be the reason?

A. It was more difficult to go through the mud pit than swim across the lake.

B. They took a lot of time to follow the hidden map and find the destination.

✓ Correct Ans **C.** They took a lot of time to swim across the lake.

D. Team B did their activities very quicker than team A.

Explanation

Team A doesn't have to swim across the lake. Therefore option C is not giving a valid reason for their loss.

Q: 5 If Sally wakes up late she will skip her breakfast and go to school.
If Sally skips her breakfast she will faint during the morning assembly.
Whenever a student faints during the assembly in the school they have to rest in the sick room.

Samuel: "If Sally faint during the morning assembly then she must have skipped her breakfast."

Raven: "If Sally rests in the sick room, then she must have woke up late."

If the information above is true, whose reasoning must be correct?

A. Samuel only

B. Raven only

C. Both Samuel and Raven

✓ Correct Ans **D.** Neither Samuel nor Raven

Explanation

If Sally skips her breakfast she will faint during the morning assembly doesn't mean whenever she faints during the morning assembly then she must have skipped her breakfast. So Samuel is wrong.
Also for the same reason, we can't certain if Sally faints during the morning assembly then she must have skipped her breakfast, so Raven is also wrong.

Q: 6 **People with choleric personalities are likely to engage in risky behaviour. However, all of them are short-tempered when they are under stress.**

Henry: "I know Ben has a choleric personality so he will not think twice to engage in risky behaviours"

Alexa: "Nyla is short-tempered when she is under stress. She must have a choleric personality."

If the information in the box is true, whose reasoning must be correct?

A. Henry only

B. Alexa only

C. Both Henry and Alexa

✓ Correct Ans **D.** Neither Henry nor Alexa.

Explanation

People with choleric personalities are likely to engage in risky behaviour. The sentence the outcome has described as a possibility. So we can't be certain that any person with a choleric personality will engage in risky behaviour. So Henry's reasoning is not correct.

All of them are short-tempered when they are under stress doesn't mean all of the people who are short-tempered when they are under stress must have choleric personalities. So Alexa is also wrong.

Q: 7 The book "Rich dad poor dad" which is written by Robert Kiyosaki and Sharon Lechter is a great book, and an inspiration for anyone who wants to improve his or her understanding of how money works.

Kim: "I read a book last year. Its cover page was damaged. At that time I was bankrupt and that book was a great inspiration for me to pull myself together and become successful"

Joseph: "Oh, then that book must be Rich dad poor dad."

Which one of these sentences shows the mistake Joseph has made?

A. Reading a book cannot help anyone to overcome the obstacles in their lives, the only solution is hard work.

B. "Rich dad poor dad" is a very old book which doesn't fit the modern financial background, so nowadays no one reads that book.

C. Joseph doesn't know about Kim's financial history.

✓ Correct Ans **D.** There might be many other books which inspire people who want to improve their understanding of how money works.

Explanation

The information given doesn't mean the book "Rich dad poor dad" which is written by Robert Kiyosaki and Sharon Lechter is the only book which inspires people who want to improve their understanding of how money works.

Q: 8 **The founder of the SGH construction company owns a privet jet and four luxurious cars.**

If the information is true, which one of the following must be true?

A. William owns a private jet and four luxurious cars, therefore he must be the founder of the SGH construction company

B. Simon owns a privet jet and also he is a founder of a company too therefore he must be the founder of SGH construction company.

✓ Correct Ans **C.** Benson owns a privet jet and only two luxurious cars, therefore he cannot be the founder of the SGH construction company.

D. Fredric has four luxurious cars. He cannot be the founder of the SGH construction company.

Explanation

Benson only owns a privet jet and two luxurious cars but the founder of the SGH construction company owns a privet jet and four luxurious cars. Therefore Benson cannot be the founder of the SGH construction company.

Q: 9 ***If Julia changes her old telephone number, her old friends will not be able to contact her via telephone. If Julia's old friends are unable to contact her, she will not get the invitation to their annual get-together. Julia will go to the get-together only if she gets the invitation. If Julia is going to the get-together she will likely buy a new gown.***

If the above information is correct, which one of the following is not possible?

A. Julia changed her old telephone number but she went to the get-together.

B. Julia got the invitation but she didn't buy a new gown.

✓ Correct Ans **C.** Julia's friends weren't able to contact her but she went to the get-together.

D. Julia didn't change her telephone number but she didn't go to the get-together.

Explanation

Even if Julia changed her number, her friends might contact her by another method instead of a telephone call. Therefore option A is possible. But if her friends weren't able to contact her she won't be able to go to the get-together. So option C is not possible.

Q: 10 *The blobfish is a pink, marine water fish that grows up to a foot long, and found in the deep waters off the coasts of Australia and Tasmania. Blobfish do not have any teeth. The blobfish is often caught in trawling nets by fishermen.*

Based on above, which one is most likely to be wrong ?

A. A fisherman found a pink fish off the coasts of Australia but it was only half a foot long therefore it can be a blobfish.

B. A fisherman found a foot-long pink fish in a river, it might be a blobfish.

C. A fisherman found a foot-long pink fish with very tiny teeth off the coasts of Tasmania, it cannot be a blobfish.

✓ Correct Ans **D.** A fisherman found a blobfish, so it must be a foot long.

Explanation

The blobfish grows up to a foot long, meaning there can be blowfishes which are smaller than that. So option D is wrong.

Q: 11 A nurse in a blood donating campaign told: "To donate blood the donator's weight must be more than 50kg and he/she be over 18 years old. You cannot donate if you have a cold, flu, sore throat, or any other infection. Also if you have recently had a tattoo or body piercing you cannot donate for 6 months from the date of the procedure."

Noah is in his late twenties. Noah does not have any piercings. Nine mounts ago he got his last tattoo.

If the above information is correct which of the following situation is a **certainty**?

A. Even though Noah is healthy and weighs more than 50kg, he cannot donate his blood because of his tattoo.

B. Noah wasn't able to donate his blood because he is too young.

C. Noah's age and weight are 27 years and 65kg. He doesn't have any health problems. He would surely not be able to donate his blood.

✓ Correct Ans **D.** Noah has a sore throat and flu today so he definitely can't donate his blood.

Explanation

The nurse insists that 'you cannot donate if you have a cold, flu, sore throat, or any other infection'. Therefore if Noah has a sore throat and flu so he definitely can't donate his blood, which is option D.

Q: 12 **Sports teacher observed:** *All the students in Adam's class are better cricket players than some students in David's class.*

Which of the following is **opposite** to sport teacher's observation ?

✓ Correct Ans **1.** There is at least one student in Adam's class who is a terrible cricket player than all of the students in David's class.

2. All the students in Adam's class are terrible cricket players.

3. All the students in David's class are better cricket players than all the students in Adam's class.

4. Students in both classes are terrible players.

Explanation

Given: all the students in Adam's class are better cricket players than some students in David's class. The negation of the above statement has the opposite meaning.
To get the negation, we have to get the negation of all the quantifiers and the statement.
Negation of "all" is "some" and negation of "some" is "all". So, the negation would be, some students in Adam's class are not better players than all the students in David's class.

Q: 13 I have opened today's newspaper and glanced through the headlines. These two caught my attention.

A: There is a low atmospheric pressure region formed.
B: The airport announced some flight cancellations.

What is the logical relation between A and B?

1. A is the effect and B is the cause.
2. ✓ Correct Ans A is the cause and B is the effect.
3. A and B are independent.
4. None of the above.

Explanation

If there is a low-pressure region in the atmosphere, it is dangerous to fly an airplane through that. That is why flights are being cancelled.

Q: 14 Emma and 9 other students are sitting in a circle participating in a game.
Starting from Emma, they are counting 1, 2, 3, 1, 2, 3, ... around the circle.
Which number will Emma say in the second round?

A. 1
✓ Correct Ans **B.** 2
C. 3
D. 4

Explanation

There are 10 students. Since 9 is a multiple of 3, 9th student will say 3 and 10th student will say 1 again. So, Emma has to say 2.

Q: 15 If the 1st of May is a Sunday, when is the last Sunday of the month?

✓ Correct Ans **A.** 29th
B. 22nd
C. 30th
D. 31st

Explanation

Last Sunday is after three or four weeks.

- After three weeks → 1 + 7 + 7 + 7 → 22nd
- After four weeks → 22 + 7 → 29th

There are more than 29 days in May. So, last Sunday is 29th.

Q: 16 Whenever Justin uses his motorcycle, Justin arrives early at school.
When Justin arrives early at school, Justin has his breakfast at the school canteen.

Logan: "Justin arrived early at school today; he must have used his motorcycle on his way here."

Paul: "Justin is having his breakfast at the school canteen; he must have used his motorcycle on his way here."

If the information is true, whose reasoning is correct?

A. Logan only

B. Paul only

C. Both Logan and Paul

✓ Correct Ans **D.** Neither Logan nor Paul

Explanation

Apply reasoning from left to right. Motorcycle -> Early at school -> Canteen. Other way round is not guranteed.

Q: 17 Whenever Joe has a runny nose, he takes anti-histamine. When he takes anti-histamine, he falls asleep.

Lester: "Joe took an anti-histamine, he must have a runny nose."

Jane: "If Joe have a runny nose then he'll fall asleep."

If the information is true, whose reasoning is correct?

A. Lester only

✓ Correct Ans **B.** Jane only

C. Both Lester and Jane

D. Neither Lester nor Jane

Explanation

Apply reasoning from left to right. Runny nose -> Anti histamine -> Fall asleep. Jane's reasoning follows left to right reasoning.

Q: 18

Tourmaline aids in the formation of a barrier around a person or space to keep bad or unwanted energies out. It is also grounding and beneficial for chakra balance. The stone can even eliminate negative thinking patterns and challenging energy, transforming them into more positive energy and beliefs. Tourmaline, as a therapeutic crystal, helps you understand yourself and others, boosts self-confidence, and reduces fear. It draws motivation, compassion, tolerance and wealth. Tourmaline, as a potent mental healer, balances the right and left hemispheres of the brain and transforms negative energy into good.

If the information is true, which of the following must be false?

A. Tourmaline is a healing crystal that fosters self-confidence, reduces fear, and aids in understanding oneself and others.

✓ Correct Ans **B.** Tourmaline stone attracts negative energy to come towards you resulting from any form of disturbance.

C. Tourmaline balances the right and left hemispheres of the brain and transforms negative energy into positive as a potent mental healer.

D. Tourmaline grounds you and aids in bringing each chakra into balance.

Explanation

Tourmaline forms a barrier around a person or space to keep away uninvited or unwanted energies out.

Q: 19

When you have anemia, your body doesn't produce enough healthy red blood cells to supply your tissues with enough oxygen. Being anemic, or having low hemoglobin, can make you feel exhausted and frail. Fatigue, weakness, pale or yellow skin, irregular heartbeats, shortness of breath, lightheadedness, chest pain, cold hands and feet, and headaches are all indications of anemia.

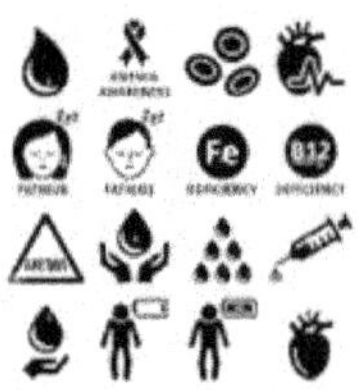

If the information is true, which of the following must be false?

✓ Correct Ans **A.** Pale skin alone is the symptoms of having an anemia.

B. Anemia is a disorder in which there are insufficient healthy red blood cells to transport oxygen to your body's tissues.

C. Anemia, commonly known as low hemoglobin, can cause fatigue and weakness.

D. Fatigue, weakness, pale or yellowish complexion, irregular heartbeats, shortness of breath, dizziness or lightheadedness are among the signs and symptoms of having an anemia.

Explanation

The information given above says that fatigue, weakness, pale or yellow skin, irregular heartbeats, shortness of breath, lightheadedness, chest pain, cold hands and feet, and headaches are all indications of anemia. Therefore, pale skin alone is the symptoms of having an anemia is false. Option A is false.

Q: 20 Those who suffer from spectrophobia may have intense phobias of their own reflection, the mirror itself, or spirits that may show up in mirrors. Despite being extremely uncommon, this condition can also be quite serious. Spectrophobia, like other phobias, can cause disruptions in many facets of a person's life and cause avoidance behaviors.

Benson: "Loisa's hair and make-up was done by her in laws. She doesn't want to even see herself in the mirror after her hair and make-up was done. She must have a spectrophobia."

Cathy: "Johan's sister doesn't like to look in the mirror. His sister fears that something might suddenly show up in the mirror. His sister may have a spectrophobia, who knows."

Whose reasoning is correct?

A. Benson

✓ Correct Ans **B.** Cathy

C. Benson and Cathy

D. Neither Benson nor Cathy

Explanation

Benson said that after Loisa's hair and make-up was done, Loisa did not want to see herself in the mirror. Benson concluded that Loisa must have a spectrophobia but there might be other reasons why Loisa doesn't want to look at herself in the mirror which makes Benson's statement to be incorrect. Cathy said that Johan's sister doesn't like to look in the mirror because his sister fears that something might suddenly show up in the mirror. Cathy concluded that Johan's sister may have a spectrophobia. Her reasoning is correct based on low modality word 'may'. Therefore, option B is the correct answer.

Q: 21 The plant compounds flavonoids, which have anti-inflammatory and antioxidant properties, are abundant in onions. These substances may offer defense against long-term diseases including cancer and diabetes when eaten often and in appropriate amounts. Eating onions can help keep your cardiovascular system in good shape. Onions contain a polyphenol called quercetin, which lowers lipid and high blood pressure levels.

If the information is true, which of the following is false?

A. Flavonoids found in onions may be a defense against cancer and diabetes.

✓ Correct Ans **B.** Onions contains polyphenol that are not good for the heart.

C. Onions must still be eaten in appropriate amounts.

D. The quercetin found in onions are found to lower lipid and high blood pressure.

Explanation

The information above says that eating onions can help keep your cardiovascular system in good shape. Onions contain a polyphenol called quercetin, which lowers lipid and high blood pressure levels. Therefore, option B is a false statement.

Q: 22 **Statement 1: "Only students who score above 85% on the physics test can attend the field trip."**
Statement 2: "Alex scored 88% on the physics test."

Natalie: "Alex will surely attend the field trip because he scored above 85% on the physics test, and only those who do can go."
Liam: "Alex can only attend the field trip if he scored above 85% on the physics test."

If the information in the bold letters is true, whose reasoning is correct?

A. Natalie Only

✓ Correct Ans **B.** Liam Only

C. Both Natalie and Liam

D. Neither Natalie nor Liam

Explanation

Natalie's reasoning is incorrect because she assumes that Alex will attend the field trip solely based on his score above 85%. However, other conditions could affect his eligibility. Liam correctly identifies that attending the field trip is contingent on scoring above 85% on the physics test.

Q: 23 A language teacher advises, "Practicing speaking with native speakers is crucial for improving language fluency."

Which one of the following statements, if true, best supports the teacher's advice?

A. Some individuals believe that they can become fluent without speaking to native speakers.

B. Finding native speakers to practice with can be challenging, especially for those in non-English-speaking countries.

C. Advanced language learning apps often replace the need for practicing with native speakers.

✓ Correct Ans **D.** Speaking with native speakers exposes language learners to authentic pronunciation and conversational nuances, contributing to fluency improvement.

Explanation

Speaking with native speakers exposes language learners to authentic pronunciation and conversational nuances, contributing to fluency improvement, which supports the teacher's advice on the importance of practicing with native speakers for language fluency.

Q: 24 **Only students who complete the advanced writing course are eligible to participate in the writing competition.**

Oliver: "Samantha didn't participate in the competition, so she must not have taken the advanced writing course."

Isabella: "Not necessarily, Samantha might have completed the course but chosen not to participate."

If the statement in the bold letters is true, whose reasoning is correct?

A. Oliver only

✓ Correct Ans **B.** Isabella only

C. Both Oliver and Isabella

D. Neither Oliver nor Isabella

Explanation

Isabella's reasoning is correct. She points out that eligibility for the competition is based on course completion, but participation is a separate decision. Samantha could have completed the course but opted not to participate in the competition.

Q: 25 **As Country X celebrates the day it gained its independence, shops, malls, and supermarkets remain closed.**

Emmy: Today isn't Independence Day because many shops are open.
Anna: Last Tuesday, I couldn't find a stall to buy groceries since everyone was closed on Independence Day.

If that's the case, who had the correct reasoning?

A. Emmy

B. Anna

✓ Correct Ans **C.** Both Emmy and Anna

D. Neither Emmy nor Anna

Explanation

Both statements are correct since shops close on Independence Day.

Q: 26 According to the results of a certain class taking Maths, English, and Crafts, the following conclusions were made :

i. *No one passed all the three tests.*
ii. *No one failed all the three tests.*
iii. *Everyone who passed Maths also passed English.*
iv. *Everyone who passed Craft also passed either Maths or English.*

So, what can't be true about Eric?

✓ Correct Ans **A.** Eric only passed Maths.

B. Eric passed both Maths and English.

C. Eric passed both Maths and Crafts.

D. Eric passed English and failed in Maths.

Explanation

A is incorrect since everyone who passed Maths also passed English. B is correct for the same reason. C is correct since one who passed Craft also passed Maths or English. D is also a probability since passing English doesn't gurantee passing Maths.

Q: 27 In a class with 4 streams (reading groups), students were asked to choose their preferred one among Neptune, Jupiter, Venus, and Pluto. Out of the 80 students, only 20 chose Neptune. The number of students who chose Jupiter was twice those who settled for Venus. Only 30 students chose Pluto.

Which of the following reasoning is correct?

A. Pluto is the least-loved stream.

B. Neptune is loved more than Jupiter.

✓ Correct Ans **C.** Venus is the least loved stream.

D. All the streams are loved equally.

Explanation

If 30 students chose Pluto, whereas 20 settled for Neptune, only 30 chose Jupiter or Venus. Since the number of Jupiter is twice that of Venus, it translates to a ratio of 2 to 1, equating to the 30 remaining students being 20 and 10, respectively. Therefore, statement A is incorrect since Venus holds that position with a following of 10 students only. On the other hand, Neptune and Jupiter tie with 20 students each; hence, B is also incorrect. C is correct due to only 10 students choosing the stream, whereas a different number of students loving various streams renders D incorrect.

Q: 28 For a team to participate in a certain tournament, its football club must meet at least one of the following conditions.

i. *Become the top 3 among the teams in its qualifying group.*
ii. *Having won at least one trophy in the last season.*
iii. *Gathering at least 20 points in the previous season.*

Which of the following teams **won't participate** in this tournament?

A. Team A, which scored 25 points during the last season and won 2 trophies

B. Team B, which was position 2 in its qualifying group but didn't get a trophy

✓ Correct Ans **C.** Team C, which hasn't won a trophy for the last 3 seasons and was position 4 in its qualifying group

D. Team D which was the winner of the Team of the Year trophy last season and gathered 30 points during that season

Explanation

Team A meets conditions 2 and 3 and qualifies to participate. Team B meets condition 1, but that's enough; hence, it qualifies for the tournament. Team C doesn't satisfy requirements, whereas Team D satisfies conditions 2 and 3. So, C is the correct answer.

Q: 29 Since the debate contest started 3 years ago, no one has ever won the trophy twice. Last year, James won the competition, whereas Richie came second. According to this year's result, James didn't break the tradition despite participating in the contest for another time. Richie also gave the contest a second shot but did even worse than he had before. The only other contestants were Janet and Elsa. If Janet took the last position, who won this time around?

A. James

B. Richie

C. Janet

✓ Correct Ans **D.** Elsa

Explanation

Since Janet was the last, she can't be the winner; hence C is incorrect. James won last year's contest but didn't break the tradition of winning the trophy once. In other words, he didn't win this time either; hence, A is also incorrect. On the other hand, Richie was second last time, and since he performed even poorer, he can't be the winner, which means B is also incorrect. Therefore, the winner was Elsa; hence, D is the correct answer.

Q: 30 There is a school trip that 40 students should attend. The teacher in charge gives chances on a **first-come, first-served** basis. The deadline for applying for this trip is Monday.

Joe: Since I submitted my application before Monday, I will surely be attending the school trip this time.
Joy: I submitted my application on Monday; hence, I won't be attending this school trip.

Given the two statements, who has the correct reasoning?

A. Both Joe and Joy

✓ Correct Ans **B.** Neither Joe nor Joy

C. Joe only

D. Joy only

Explanation

Whereas Joe submitted his application before Monday, the first come, first served rule also applies, which means that's not a guarantee of going on this school trip. Equally important, just because Joy submitted hers on the last doesn't mean she wasn't among the first 4o students who applied for the school trip. So, both reasonings are incorrect, and B is the correct answer.

Q: 31 John and Mary like communicating using a code. Therefore, they have devised a way of decoding messages to maintain privacy. The code is as follows:

A	B	C	D	E	F	G	H	I	J	K	L	M	N	O	P	Q	R	S	T	U	V	W	X	Y	Z
1	2	3	4	5	6	7	8	9	0	1	2	3	4	5	6	7	8	9	0	1	2	3	4	5	6

Under such circumstances, which words can share the same code as BOOK?

✓ Correct Ans **A.** LOOK

B. VASE

C. LYOD

D. BELL

Explanation

According to the code, BOOK translates to 2-5-5-1, which applies to LOOK. On the other hand, VASE is 2-1-9-5, LYOD is 2-5-5-4, and BELL is 2-5-2-2. So, A is the correct answer.

Q: 32 From a certain location, Agnes walks 70 m towards the south. She then turns right and travels a distance of 70 m straight. She turns left, walking up to 60 m straight. Lastly, she turns left and walks for 70 m straight. So, what's the distance between where she is and where her journey began?

A. 270 m

B. 200 m

✓ Correct Ans **C.** 130 m

D. 70 m

Explanation

As shown in the diagram below, there is a distance of 60 m and another one that's 70 m between where Agnes was and the initial point. So, the distance is 70 + 60 = 130 m.

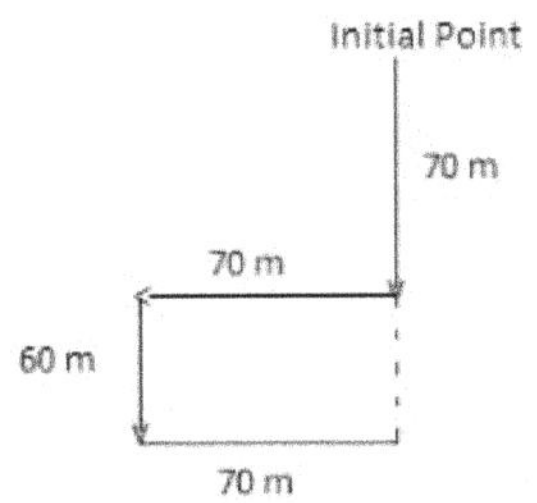

Q: 33 **If you walk regularly, going to the gym to work out is unnecessary. Equally important, you don't need fitness equipment to exercise. The only thing necessary is comfortable shoes, especially athletic ones.**

Which of the following is the best **interpretation** of the above statements?

A. Walking regularly is better than working out.

B. If you don't use athletic shoes, you will hurt your feet when walking.

✓ Correct Ans **C.** Walking regularly is as effective as working out to keep fit.

D. You should combine walking and working out for fitness.

Explanation

The paragraph suggests walking regularly and working out as equivalent exercises; hence, A and D are wrong, whereas C is correct. Regarding the shoes, comfort matters the most, and the athletic option is more of a recommendation; hence, B is also incorrect reasoning.

Q: 34 It is the year 2000, and Alex has been asked by his class teacher to calculate the number of days in the months with an even number of days. He follows the following steps.

i. **Out of the months with an even number of days, there are February, April, June, September, and November.**
ii. **The number of days are 28, 30, 30, 30 and 30.**
iii. **The sum is 28 + 30 + 30 + 30 + 30=148**

Which of the following is conclusion regarding Alex's approach ?

A. Alex got it right.

✓ Correct Ans B. Alex was bound to fail solving the question after step 1 because step 1 is incorrect.

C. Alex was bound to fail solving the question after step 2 because spep 2 is incorrect.

D. Alex was bound to fail solving the question after step 3 because step 3 is incorrect.

Explanation

Since it is 2000, that's a leap year. Hence February has 29 days which isn't even.

Q: 35 A train from Town A to B picks 4 people along the road.

i. **Jane isn't the first or the last person to board to train.**
ii. **Only Emily enters after Jane.**
iii. **Agatha enters the train first.**

If the first two statements are true, the third one is ....

A. True

B. False

✓ Correct Ans C. Uncertain

D. None of the above

Explanation

According to the first and the 2nd statements, Jane->Emily. Agatha must be one of the other two people besides Jane and Emily who board the train. It is hard to tell whether she is the first one to board and it is the other individual. So, C is correct.

Q: 36 If the exchange rate of USD vs Euro is 1.2 USD for 1 Euro, which of the following reasoning is correct?

Mark: If I exchange my USD for Euros, the number of Euros will be relatively high.
Jimmy: Since I will be exchanging my Euros for USD, the USD will be relatively lower.

A. Mark

B. Jimmy

C. Both Mark and Jimmy

✓ Correct Ans **D.** Neither Mark nor Jimmy

Explanation

Since the value of Euro is higher than that of USD, you will need more USD than Euros to have the same value. Therefore, both reasons are incorrect.

Q: 37 Different people had various opinions regarding Daniel's weight. **A** said that the weight was between 60 and 70 kg. **B** had a different range of between 50 and 65 kg. **C** said that the weight is less than or equal to 62.
If all the opinions were factual, which of the following conclusions is correct?

A. Daniel's weight can be 65

B. Daniel's weight can also be 68

C. Daniel's weight can be 58

✓ Correct Ans **D.** Daniel's weight can be 60

Explanation

First, all arguments are correct. So, according to A, the weight is between 60 and 70. Given B's range, the extreme range drops from 70 to 65. So, the weight is between 60 and 65. C lowers the upper range to 62 from the previous 65. So, Daniels's weight can neither be 65 nor 68. A and B are incorrect. On the other hand, it can't be 58 since that's lower than the 60 cap. So, D is correct since 60 is between 60 and 62.

Q: 38 In an exam no two students got the same marks. If you want to find out one student's rank (from top) in the exam, which of the following statements would help you figure it out?

i. ***The class has 50 students.***
ii. ***Only 15 students scored more than the individual in question.***

A. i only

✓ Correct Ans **B.** ii only

C. both statements together

D. none of the above

Explanation

The total number of students isn't necessary for figuring out how one has ranked. However, knowing how many people have done better than the person is necessary, hence B is correct.

Q: 39 There are three sentences :

Jake read more books than Antonio.
Antonio read fewer books than Abby.
Abby read more books than Jake.

If the first two statements are true, the third one is....

A. True

B. False

✓ Correct Ans **C.** Insufficient data to connect

D. None of the above

Explanation

The only thing one can deduce from the first two statements that are true is that Jake and Abby read more books than Antonio. However, one can't tell with certainty who between Jake and Abby has read more than the other.

Q: 40 To avoid idleness in the office, four workers have decided to work in shifts, each working alone in a day, from Monday to Thursday.
The schedule showed that Job (office worker) was to work on Monday. However, he devised an arrangement to switch dates with Leon, who was to work on Wednesday. Ann also traded her working day with Emily, who (Emily) was to work on Thursday.

So, who was on duty on Tuesday, eventually?

A. Job

✓ Correct Ans **B.** Emily

C. Ann

D. Leon

Explanation

According to the schedule, Job was to work on Monday, Leon on Wednesday, and Emily on Thursday, thus leaving Tuesday for Ann. However, Ann and Emily changed their working arrangement; hence, Emily worked on Tuesday instead, and B is the correct answer.

Answers And Solutions
Learnikx Education

Q: 1 Who is L. Frank Baum?

A. The character in The Wonderful Wizard of Oz children's novel
B. The director of The Wonderful Wizard of Oz movies
✓ Correct Ans **C.** The author of The Wonderful Wizard of Oz children's novel
D. One of a tornado victim

Explanation

L. Frank Baum is the one who wrote the story. So the answer is C.

Q: 2 What is the other name of a tornado?

✓ Correct Ans **A.** Twisters
B. Toaster
C. Rotating winds
D. Thunder

Explanation

According to the article, "twisters" is another term for tornedos. So the answer is A.

Q: 3 Based on the article above, Kansas is known for many things except ...

A. Tornadoes
✓ Correct Ans **B.** Tomatoes
C. Wheat
D. Sunflowers

Explanation

At the start of the article, it is stated that Kansas is known for many things -- wheat, sunflowers ... and tornadoes! There are no tomatoes mention in the article.

Q: 4 The winds that rotate within a tornado usually reach a speed of ...

✓ Correct Ans **A.** Almost 300 miles per hour

B. 300-350 miles per hour

C. More than 300 miles per hour

D. None of the above

Explanation

In the third paragraph, the writer mention that a tornado usually reach a speed of almost 300 miles per hour. So the answer is A.

Q: 5 When did the Romans finish building the Coliseum?

A. The year 70

B. The year 523

✓ Correct Ans **C.** The year 80

D. The year 240

Explanation

According to the 1st paragraph, they started building it in the year 70. It took 10 years to construct. Option C is correct.

Q: 6 What caused the fire that damaged the upper levels of the Coliseum?

✓ Correct Ans **A.** A bolt of lightning

B. Rowdy people who came to watch the events

C. An attacking army

D. An angry mob

Explanation

The coiseum was struck by lightning in the year 217. This started a fire. Option A is supported.

Q: 7 Which caused the most damage to the Coliseum?

- **A.** Fires
- ✓ Correct Ans **B.** Earthquakes
- **C.** Wars
- **D.** Hurricanes

Explanation

According to the 4th paragraph, the worst damage happened in 1349. A mighty earthquake shook Rome and the Coliseum. This supports option B.

Q: 8 Both extracts discuss on ..

- **A.** Abdominal pain
- ✓ Correct Ans **B.** Severe headache
- **C.** Brain tumor
- **D.** Fashion

Explanation

Migraine is a type of severe headache. Both extracts discuss migraine.

Q: 9 What is an aura in relation to a migraine?

- **A.** A medicine that can prevent or lessen migraines
- **B.** A severe throbbing pain or pulsing sensation in the head
- ✓ Correct Ans **C.** A warning symptom that can occur before or during migraines
- **D.** None of the above

Explanation

An aura can include visual disturbances or tingling sensations that can happen before or with a migraine headache.

Q: 10 What might make the pain of a migraine return even after the initial attack has ended?

- ✓ Correct Ans **A.** Sudden head movements
- **B.** Eating spicy food
- **C.** Drinking caffeine
- **D.** None of the above

Explanation

Sudden head movements might bring on the pain again briefly even after the migraine attack has ended.

Q: 11 How does ginger help with migraines?

A. Ginger can relieve dehydration caused by migraines.

✓ Correct Ans **B.** Ginger contains phytochemicals that can manage headaches.

C. Ginger can be made into a vaporizer for inhalation.

D. Ginger has been proven to be an alternative to modern medicine.

Explanation

In Extract B, there are indications that ginger has phytochemicals that might be responsible for effectively managing headaches.

Q: 12 Why should home remedies for migraines be used cautiously?

A. Home remedies are not effective in managing migraine symptoms.

B. Home remedies can cause more harm than good.

✓ Correct Ans **C.** The benefits of home remedies have not been scientifically proven.

D. Home remedies should only be used as a substitute for modern medicine.

Explanation

According to Extract B, while some studies show the benefits of the given herbs and home remedies for migraines, large-scale human research is necessary to understand their extent of benefits on human health. Therefore, they should only be taken cautiously and never as a substitute for medical treatment.

Q: 13 Which of the extracts refer to the harbor front Sydney opera house?

A. Extract A

B. Extract B

C. Extract C

✓ Correct Ans **D.** Extract D

Explanation

Extract D states that Sydney is best known for its harbor front Sydney Opera House.

Q: 14 Which of the extracts describe context of stunning beaches, vibrant food scene, and culture?

A. Extract A

B. Extract B

C. Extract C

✓ Correct Ans **D.** Extract D

Explanation

According to Extract D, Sydney, with its stunning beaches, vibrant food scene, and lively arts and culture, is a city that never fails to captivate.

Q: 15 Which of the extracts mention Robben island?

✓ Correct Ans **A.** Extract A

B. Extract B

C. Extract C

D. Extract D

Explanation

Extract A describes about the busy harbor and boats heading for Robben Island.

Q: 16 Which one fits in (1) ?

A. Sentence A

B. Sentence B

✓ Correct Ans **C.** Sentence G

D. Sentence H

Explanation

The essay in on the doctor-patient relationship. Therefore, the correct sentence should be a introduction to it.

Q: 17 Which one fits in (2) ?

✓ Correct Ans **A.** Sentence A

B. Sentence B

C. Sentence C

D. Sentence D

Explanation

The next line states about the trust and the correct answer should have the word “trust in it”.

Q: 18 Which one fits in (3) ?

A. Sentence A

B. Sentence D

✓ Correct Ans **C.** Sentence B

D. Sentence E

Explanation

The paragraph is based on the trust. The correct answer should be a elaboration on the trust and its outcomes.

Q: 19 Which one fits in (4) ?

A. Sentence A

B. Sentence B

C. Sentence H

✓ Correct Ans **D.** Sentence E

Explanation

The word "compassion" should be included in the correct answer because the paragraph starts with that word.

Q: 20 Which one fits in (5) ?

A. Sentence A

B. Sentence B

✓ Correct Ans **C.** Sentence C

D. Sentence D

Explanation

The paragraph explains about the difference between the doctor patient relationship in the past and present. Thus, the correct answer should be about the health care partnership in contrast to the past.

Q: 21 Which one fits in (6) ?

A. Sentence A

B. Sentence B

C. Sentence C

✓ Correct Ans **D.** Sentence D

Explanation

The next parts of the paragraph explain about the advantages of a healthy doctor patient relationship. Therefore, the first line should be an introductory to that.

Q: 22 Which one fits in (7) ?

A. Sentence A

B. Sentence G

✓ Correct Ans **C.** Sentence H

D. Sentence C

Explanation

The correct answer is the first line of the last paragraph, thus, ideally, the word “in conclusion” should be inserted.

Q: 23 In the poem, what does the phrase "the daily race" (line 2) most likely refer to?

A. A marathon

B. A competitive sport

✓ Correct Ans **C.** The challenges of daily life and work

D. A horse race

Explanation

The phrase "the daily race" in line 2 most likely refers to the challenges and demands of daily life and work.

Q: 24 What is the deeper meaning conveyed by the line "But glad in his heart that his own rejoice" ?

✓ Correct Ans **A.** The father finds joy in his children's happiness.

B. The father is happy because he has achieved fame.

C. The father is glad to be done with his daily work.

D. The father is unhappy despite his children's joy.

Explanation

The line suggests that the father's joy comes from seeing his children happy and welcoming him home.

Q: 25 What does the poet mean by "The deeds that his father did for him" ?

✓ Correct Ans **A.** The father's actions and sacrifices for his children

B. The father's achievements and successes

C. The father's hobbies and interests

D. The father's mistakes and failures

Explanation

The phrase "The deeds that his father did for him" refers to the actions and sacrifices made by the father for his children.

Q: 26 What is the speaker's attitude towards the father in the poem?

✓ Correct Ans **A.** Admiration

B. Indifference

C. Criticism

D. Envy

Explanation

The speaker's attitude towards the father is one of admiration for his sacrifices and selflessness.

Q: 27 What is the central message or theme of the poem "Only a Dad"?

A. Fathers are unimportant in the lives of their children.

✓ Correct Ans **B.** The sacrifices of fathers often go unnoticed.

C. Fathers should focus on achieving fame and wealth.

D. Children should express their gratitude to their fathers regularly.

Explanation

The central theme of the poem is that the sacrifices and efforts of fathers often go unnoticed or unappreciated.

Q: 28 Which line in the poem uses metaphor to describe nature's initial state?

✓ Correct Ans **A.** "Nature's first green is gold"

B. "Her hardest hue to hold"

C. "So dawn goes down to day"

D. "Nothing gold can stay"

Explanation

Line 1, "Nature's first green is gold," uses metaphor to compare the initial green of nature to the color gold.

Q: 29 How does the poet use sound alliteration in the poem?

A. "But only so an hour."

B. "So Eden sank to grief,"

C. "Nothing gold can stay."

✓ Correct Ans **D.** "Her hardest hue to hold."

Explanation

The line "Her hardest hue to hold" uses alliteration with the repeated "h" sound.

Q: 30 What deeper meaning or message does the poem convey about life?

A. Life is eternal and unchanging.

B. Life is filled with constant growth and renewal.

✓ Correct Ans **C.** Life is brief and transient, and beauty fades.

D. Life is unpredictable and chaotic.

Explanation

The poem conveys the deeper message that life is brief and transient, and beauty does not last forever.

Q: 1 How many possible combinations of 1 pencil and 1 rubber cap can be formed from 3 pencils and 4 rubber caps?

A. 7

B. 10

✓ Correct Ans **C.** 12

D. 14

Explanation

This is picking problem. 3x4 i.e. 12 combinations possible.

Q: 2 Find the hidden numbers to get the same value when the numbers in the boxes below are added across and down.

?	4	6
?	6	5
?	8	7

A. 2, 3, 6

✓ Correct Ans **B.** 8, 7, 3

C. 6, 2, 3

D. 7, 6, 2

Explanation

Add the numbers in the boxes below and find the value.

4+6+8=18 and 6+5+7=18

So we know that the value is 18. Let's add the given numbers and find out how many should be added to 18.

4 +6 =10, 6+5=11, 8+7=15
18-10=8, 18-11=7, 18-15=3

The correct answer is: B. 8, 7, 3

Q: 3 According to the pattern, what should be the numbers for A and B respectively?

2	3	4	24
1	A	3	15
4	2	B	32
8	30	48	

A. 6 and 3

B. 3 and 3

✓ Correct Ans **C.** 5 and 4

D. 4 and 3

Explanation

Let's find out what pattern the surrounding numbers are made of.
$24 = 2 \times 3 \times 4$, $8 = 2 \times 1 \times 4$
$1 \times A \times 3 = 15$, $3A = 15$ (Divide by 3) $A = 5$
$30 = 3 \times A \times 2$, $30 = 3 \times 5 \times 2$ (By substituting $A=5$)
$4 \times 2 \times B = 32$, $8B = 32$ (Divide by 8) $B = 4$
$4 \times 3 \times B = 48$, $4 \times 3 \times 4 = 48$ (By substituting $B = 4$)
The correct answer is: C. 5 and 4

Q: 4 Ricky paid \$473 for 5 similar cushions and 3 similar blankets. Shawn paid \$339 for 3 similar cushions and 3 similar blankets.
How much is the cost of 3 similar cushions?

✓ Correct Ans **A.** 201

B. 302

C. 134

D. 339

Explanation

Let us write the equations and then subtract.

5C + 3B = 473
<u>3C + 3B = 339</u>
2C + 0 = 134

C = 134 ÷ 2 = 67
3C = 67 X 3 = \$201

The cost of 3 similar cushions is \$201.

Q: 5 Namo saved \$30 more than Chitra. Aria saved 3 times as much money as Namo. The 3 girls saved a total of \$300. How much money did Namo save?

A. 22

B. 23

C. 33

✓ Correct Ans **D.** 66

Explanation

Namo = n
Chitra = n-30
Aria = 3n
Total = n + (n-30) + 3n = 5n - 30 = 300;
5n = 300+30 = 330; n = 66

Q: 6 Aston has $180 more than Bob. Kevin has twice as much as Aston. The three children have a total amount of $1140. How much does Bob have?

A. 110

B. 120

C. 140

✓ Correct Ans **D.** 150

Explanation

Bob represented by n, Aston by n + 180, Kevin by 2n + 360.

n + n + 2n + 540 = 1140
4n = 600
n = 150

Q: 7 There were 38 children at a fun-fair. Every boy received 2 balloons each and every girl received 4 balloons each at the fun-fair. A total of 126 balloons were given to the children.
How many boys were there at the fun-fair?

A. 10

B. 11

C. 12

✓ Correct Ans **D.** 13

Explanation

There were 'n' boys and '38-n' girls.
n x 2 + (38-n) x 4 = 126
38x4 - nx2 = 126; nx2 = 152 - 126 = 26
n = 13

Q: 8 Amirah had a collection of 20 old coins. The value of the coins was $6.40 in total.
In her collection, there were only 20-cent coins and 50-cent coins.
How many 50-cent coins did she have?

A. 6

B. 7

C. 9

✓ Correct Ans **D.** 8

Explanation

'n' 50 cent coins.
'(20-n)' 20 cent coins.

Total value :
50n + (20-n)x20 = 640
50n -20n = 640 - 400 = 240
30n = 240; n = 8

Q: 9

Lynn had 93 pencils and pens. After giving away $\frac{3}{4}$ of her pens and 28 pencils, she had an equal number of pencils and pens left.
How many pencils did she have at first?

A. 65

B. 42

✓ Correct Ans **C.** 41

D. 40

Explanation

Start : Pen = n, Pencil = 93 - n
End : Pen = n/4, Pencil = 65 - n
n/4 = 65 - n; n/4 + n = 65
n x 5/4 = 65 ; n = 65 x (4/5) = 52
Pencil = 93 - n = 93 - 52 = 41

Q: 10

Lucas had some sweets.
If he gave 7 sweets to each of his friends, he would be short of 4 sweets. If he gave 6 sweets to each of them, he would have 3 sweets left.

How many sweets did Lucas have?

A. 14

✓ Correct Ans **B.** 45

C. 17

D. 7

Explanation

Lucas had 'n' friends.
nx7 - 4 = nx6 + 3
n = 7

Lucas had = nx7 - 4 = 49 - 4 = 45 sweets

Q: 11 Jiemin has some orange juice. If he fills 20 identical cups with the orange juice, he will have 600 ml of the orange juice left. If he tries to fill 6 identical jugs with the orange juice instead, he will be short of 400 ml of the orange juice. The capacity of a jug is the same as that of 4 such cups.
What is the capacity of a jug?

A. 1500ml

B. 2000ml

C. 5000ml

✓ Correct Ans **D.** 1000ml

Explanation

Capacity of 6 jugs = capacity of 24 cups
To fill 20 cups → 600 ml left
To fill 24 cups → short of 400 ml
Capacity of 4 caps → 600 + 400 =1000 ml
Capacity of 1 jug = Capacity of 4 cups = 1000ml

Q: 12 A drink stall owner sold $\frac{1}{9}$ of his canned drinks in the morning and $\frac{3}{4}$ of the remaining canned drinks in the afternoon. After he received a delivery of another 200 canned drinks, he found that he had 32 canned drinks more than what he had at first. How many canned drinks did the stall owner have at first?

✓ Correct Ans **A.** 216

B. 215

C. 168

D. 24

Explanation

200 - 32 = 168

Morning = $^1/_9$
Remaining = $^8/_9$
Afternoon = $^8/_9 \times {}^3/_4 = {}^{24}/_{36} = {}^2/_3$

Total = $^1/_9 + {}^2/_3 = {}^7/_9$

$^7/_9$ = 168; Whole = 216

Q: 13 A fruit stall sells apples, oranges and pears. The ratio of the number of pears to the total number of fruits is 2 : 5. There are 300 apples and 180 oranges.

Find the number of pears at the stall.

A. 360 pears

B. 160 pears

✓ Correct Ans **C.** 320 pears

D. 480 pears

Explanation

Pear : All fruits = 2 : 5

5u - 2u = 3u
3u = 300 + 180 = 480
1u = 160; 2u = 160 X 2 = 320

Q: 14 The bar chart below shows the number of sales of 4 types of animals at a pet store.

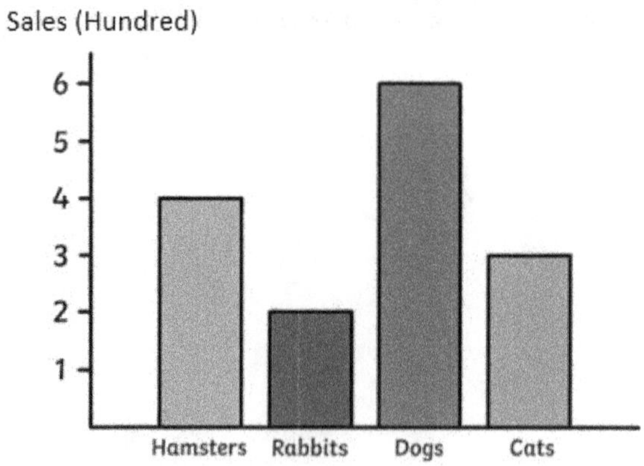

Which of the following is a correct statement?

A. The sales of hamsters are the highest.

✓ Correct Ans **B.** The difference in sales between dogs and cats is 300.

C. The sum of sales of rabbits and hamster is 500.

D. None of the above.

Explanation

A. The sales of hamsters are the highest, wrong statement as the dog is the highest.
B. The difference in sales between dogs and cats is 300, correct statement, 600 dogs – 300 cats = 300
C. The sum of sales of rabbits and hamster is 500, wrong statement, 400 hamsters + 200 rabbits = 600
D. None of the above is wrong as option B is correct

Q: 15 Diagram below shows a triangle

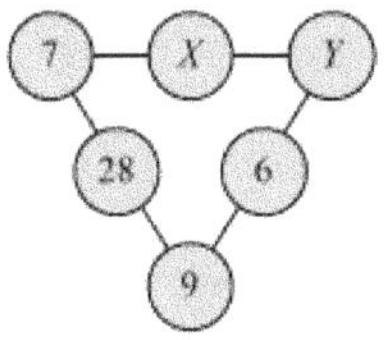

The total of the 3 numbers on each side of the triangle is the same. What are the values of X and Y?

A. X = 8, Y = 39

B. X = 7, Y = 29

✓ Correct Ans **C.** X = 8, Y = 29

D. X = 10, Y = 29

Explanation

7 + 8 + 9 = 44
9 + 6 + Y = 44
Y = 29
7 + X + Y = 44
X = 44 – 29 – 7 = 8

Q: 16 The graph shows the number of ice creams sold.

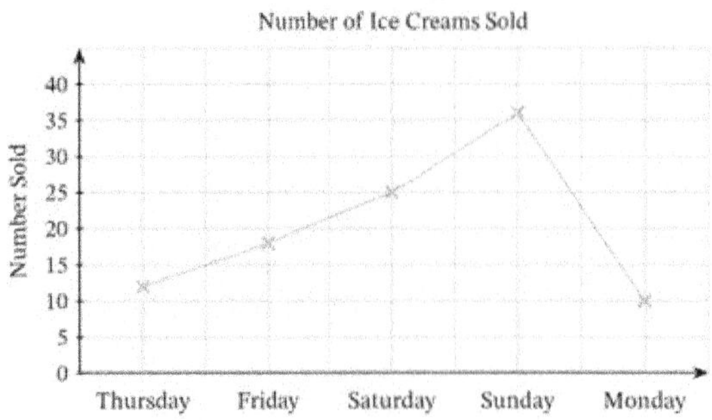

Which of the following is correct?

A. The highest sales are on Saturday

✓ Correct Ans **B.** The number of ice creams sold on Saturday and Monday is 35.

C. The lowest sales are on Sunday.

D. None of the above.

Explanation

A. The highest sales are on Saturday. Wrong statement as Sunday has the highest sales.

B. The number of ice creams sold on Saturday and Monday is 35. Correct statement, Saturday, 25 plus Monday, 10 = 35

C. The lowest sales are on Sunday. Wrong statement as Monday has the lowest sales.

D. None of the above. Wrong statement as option B is correct

Q: 17 The graph below shows the temperature on 1st September 2022.

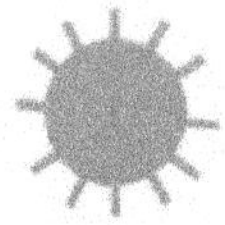

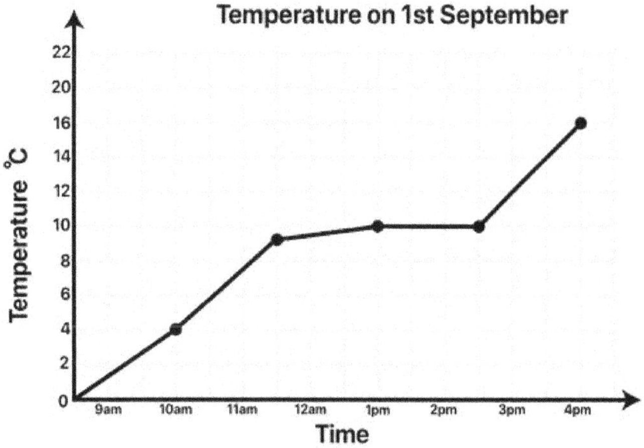

Which of the followings is a wrong statement based on the graph?

A. It is 16°C at 4 in the evening

B. It is 10°C at 1 in the afternoon

✓ Correct Ans **C.** It is 4°C at 11 in the morning

D. The highest temperature recorded is at 4 pm

Explanation

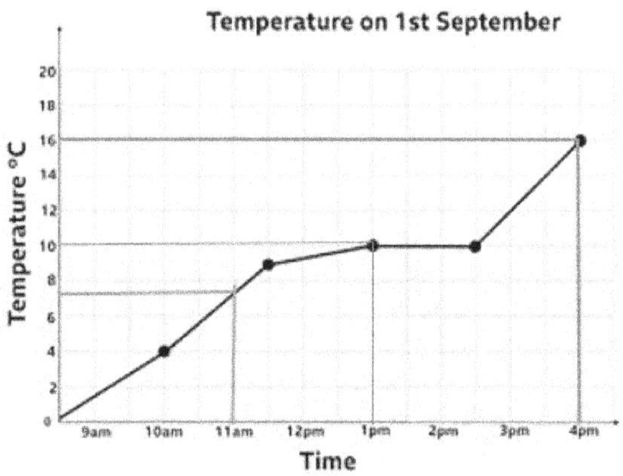

Option A and option D are correct
Option B is correct
Option C is wrong

Q: 18 The table shows the number of beads in a container.

Color	Number Of Beads
White	50
Red	30
Blue	$\frac{2}{5}$ of the number of white beads

Daisy uses $^3/_5$ white beads and 40% of red beads to embroider her dress.
Count the number of beads left in the container.

A. 25

B. 84

✓ Correct Ans **C.** 58

D. 75

Explanation

Blue = $^2/_5$ x 50 = 20

White used = $^3/_5$ x 50 = 30, balance 20

Red used = 30 x 40% = 12, balance 18

Remaining: 20 + 20 + 18 = 58

Q: 19 Below shows a dialogue between four students who collect donations.

Larry	: I was able to collect $2.50.
Mei Ling	: The amount of money I collected was 90 cents more than Larry.
Raj	: My collection money is the same as Mei Ling's money.
John	: If added to the money I collected, the total is $12.

Based on the dialogue, count the money that John successfully collected.

✓ Correct Ans **A.** $2.70

B. $3.70

C. $4.50

D. $2.30

Explanation

Larry : 2.50

Mei : 2.50 + 0.90 = 3.40

Raj : 3.40

Total = 2.50 + 3.40 + 3.40 = 9.30

John = 12 – 9.30 = 2.70

Q: 20 Rayyan sets an alarm at 5.40 a.m. When the alarm rings, he pressed the snooze button and goes back to sleep. The alarm rang every 5 minutes. If Rayyan wakes up at 6.10 a.m. when alarm rang, how many times does the alarm ring?

A. 5

B. 6

C. 10

✓ Correct Ans **D.** 7

Explanation

5.40 until 6.10 = 30 minutes he took to wake up
30 ÷ 5 = 6
Final alarm 1 time. Total 7 times.

Q: 21 At a party, each girl was given 3 balloons and each boy was given 5 balloons. There were 32 children at the party and a total of 116 balloons were given out. How many boys were there at the party?

✓ Correct Ans **A.** 10

B. 11

C. 3

D. 8

Explanation

Assume all are girls → 32 x 3 = 96 balloons
Total Difference → 116 - 96 = 20 balloons
Difference between 1 boy and 1 girl → 5 - 3 = 2 balloons
Number of boys = 20 ÷ 2 = 10

Q: 22 Figure A is made up of a square and 8 identical triangles. Each triangle has a height of 7 cm and a base of 3.5 cm. Find the area of Figure A.

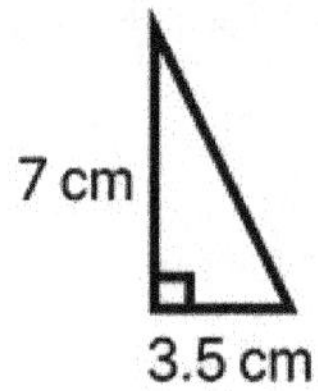

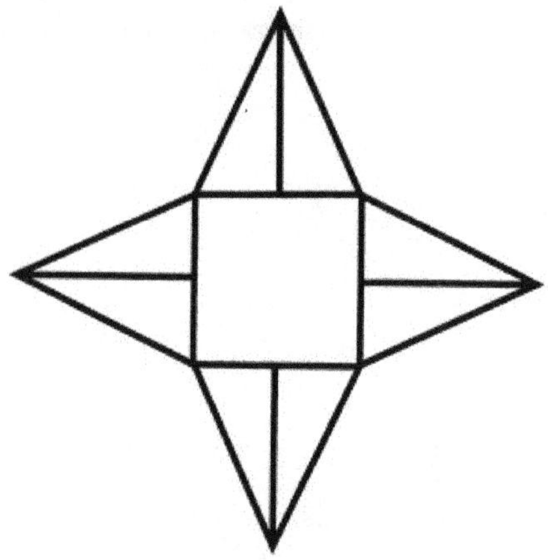

Figure A

A. 124 cm^2

B. 139 cm^2

C. 149 cm^2

✓ Correct Ans **D.** 147 cm^2

Explanation

Area of 1 triangle → $^1/_2$ x 3.5 x 7 = 12 $^1/_4$
Area of 8 triangles → 12 ¼ x 8 = 98
One side of the square =3.5 +3.5 = 7 cm
Area of the square = 7x7 cm^2

98 + 49 = 147 cm^2

Q: 23 If the length of the side of a square is doubled, what is the ratio of the areas of the original square to the area of the new square?

A. 1:5

✓ Correct Ans **B.** 1:4

C. 1:3

D. 1:9

Explanation

If x is the side of the original square, then its area is equal to x^2
If x is doubled to 2x, then the new area is equal to $(2x)^2 = 4x^2$
The ratio of the areas of the original square to the area of the new square = $x^2 / (4x^2)$ = 1/4 or 1:4

Q: 24 In the figures given below, ABCD is a rhombus.

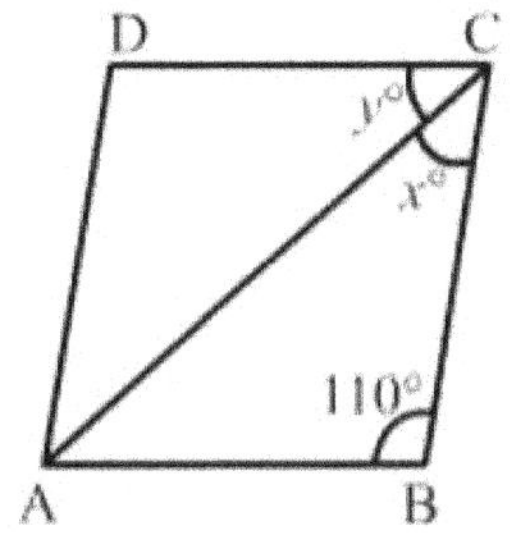

Find the value of x and y.

✓ Correct Ans **A.** x = 35°, y=35°

B. x = 45°, y = 35°

C. x = 55°, y = 45°

D. x = 35°, y = 65°

Explanation

ABCD is a rhombus. A rhombus has four equal sides.
In triangle ABC, angle BAC = angle BCA
x = ½ (180-110) = 35°

Same calculation for y.

Q: 25 The ratio of the number of boys and girls in a college is 7: 8. If the percentage increase in the number of boys and girls is 20% and 10% respectively, what will be the new ratio?

A. 8:9

B. 17:18

✓ Correct Ans **C.** 21:22

D. Cannot be determined

Explanation

Boys = 7 parts, Girls = 8 parts
Boys = 8.4, Girls = 8.8 parts
New ratio : 8.4 : 8.8 = 84:88 = 21:22

Q: 26

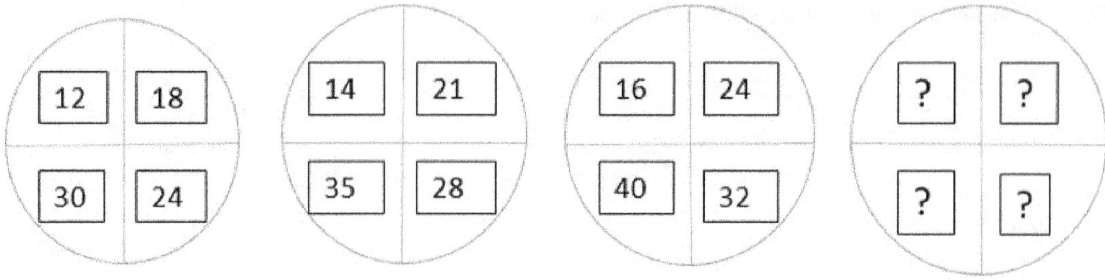

What are the 4 numbers that are suitable for the quarters of the 4 circles?

A. 22,28,34,45

✓ Correct Ans **B.** 18,27,36,45

C. 24,32,42,54

D. 26,35,40,58

Explanation

First 3 circles contain multiples of 6, 7, and 8 respectively. Accordingly 4th circle should contain multiples of 9. Answer b has multiples of 9.

Q: 27 There are white keys in a small key board that are named as C, D, E, F, G, A and B. What is the probability of pressing E note by a blind folded person in his all 2 chances?

A. $^1/_7$

B. $^2/_7$

✓ Correct Ans **C.** $^1/_{49}$

D. $^2/_{49}$

Explanation

There are 7 keys.
Probability of pressing E note in 1st time= $^1/_7$
Probability of pressing E note in 2nd time =$^1/_7$
Probability of pressing E note in both times= $^1/_7 \times {}^1/_7 = {}^1/_{49}$

Q: 28 Simplify and find the value in scientific notation.

$6.78 \times 10^8 + 17.5 \times 10^7$

A. 24.28×10^{15}

B. 24.28×10^7

C. 8.53×10^{16}

✓ Correct Ans **D.** 8.53×10^8

Explanation

When adding values with powers we have to convert the powers in to a same number. Let's convert power 7 into power 8.

$17.5 \times 10^7 = 1.75 \times 10^8$

Then we can add them. When adding, add numbers without powers and put the power after that sum.

Here, add 6.78 and 1.75. Answer is 8.53 then put 10^8 after that. We get 8.53×10^8 as the final answer.

Q: 29 Which of the below options is correct?

A box contains 9 sprite bottles, 18 Coca-Cola bottles and 27 Fanta bottles.

The ratio 1:6 compares to ...

A. None of the options

B. Sprite bottle to Fanta bottles

C. Fanta bottles to Coca-Cola bottles

✓ Correct Ans **D.** Sprite bottles to all the bottles

Explanation

Ratio of sprite bottles to Coca-Cola bottles to Fanta bottles is 9:18:27. All the numbers are divisible by 9. When the above numbers are divided by 9, the ratio is 1:2:3. Ratio 1:6 is equivalent to no. of sprite bottles to total no. of bottles in the box.

Q: 30 Aina is a worker in a shoe shop. She has 5 types of shoes to keep in racks. How many different ways can she arrange these shoes in a row?

A. 5

B. 25

C. 60

✓ Correct Ans **D.** 120

Explanation

This is 'arrange type' question. Answer is factorial 5.

$= 5 \times 4 \times 3 \times 2 \times 1$

$=120$

Q: 31 John had 4 times as many stickers as Mark. Dan had 2 times as many stickers as John. Mark had 420 stickers fewer than John. How many stickers did Dan have?

A. 1440

B. 1560

✓ Correct Ans **C.** 1120

D. 1020

Explanation

Mark = n; John = nx4; Dan = nx8
nx4 - n = 420; nx3 = 420; n = 140
Dan = nx8 = 140x8 = 1120

Q: 32 Four children shared some sweets among themselves. Adam took $^1/_8$ of the total number of sweets while Barry took $^1/_4$ of them. Kaden took $^1/_4$ of the remaining sweets and Shaun took the rest.
If Barry took 20 more sweets than Adam, how many sweets did Shaun take?

✓ Correct Ans **A.** 75

B. 160

C. 25

D. 100

Explanation

$^1/_4 = ^2/_8$
$^2/_8 - ^1/_8 = ^1/_8$
$^1/_8$ of total = 20
Adam took 20 sweets.
Barry → 20 + 20 = 40
Total sweets → 8 × 20 = 160
Remaining sweets → 160 – 40 – 20 = 100
Kaden → $^1/_4$ × 100 = 25
Shaun → 100 – 25 = 75
Shaun took 75 sweets.

Q: 33 A train carrying 72 adults left Stop X for Stop Y.

$^{6}/_{11}$ of the passengers were adults and the rest were children.

At Stop Y, $^{1}/_{4}$ of the children got off the train.

How many children were on the train when the train left Stop Y?

A. 25

B. 15

✓ Correct Ans **C.** 45

D. 60

Explanation

$^{6}/_{11}$ of passengers → 72

$^{1}/_{11}$ of passengers → 72 ÷ 6 = 12

$1 - {}^{6}/_{11} = {}^{11}/_{11} - {}^{6}/_{11} = {}^{5}/_{11}$

5/11 of passengers (children) → 5 × 12 = 60

Children who got off → 1/4 × 60 = 15

Children who remained → 60 – 15 = 45

45 children were on the train.

Q: 34 The table below shows the different types of fruits Mr. Malhotra sold.

Fruits	Number of fruits sold
Oranges	50
Pears	30
Mangoes	
Avocadoes	20

The number of mangoes sold was $^{1}/_{5}$ of the total number of fruits sold. How many mangoes were sold?

A. 30

B. 35

C. 20

✓ Correct Ans **D.** 25

Explanation

Number of oranges, pears, and avocadoes sold → 50 + 30 + 20 = 100

4/5 of fruits sold → 100

1/5 of fruits sold → 100 ÷ 4 = 25

25 mangoes were sold.

Q: 35 A box of milk has 2.5 L milk. Jenny drank half of the milk. Lisa drank 20% of the milk left. Rosie put another 190ml of milk into the box.
How much milk is in the box now?

✓ Correct Ans **A.** 1,190ml

B. 1,458ml

C. 1,100ml

D. 1,300ml

Explanation

2.5ℓ = 2500ml
2500ml ÷ 2 = 1250ml x 20% = 250
1250 – 250 = 1000 + 190 = 1190

Q: 1 Read the statements and decide the validity of conclusion.
Note : For the purpose of this question, statements are always to be taken as correct, even if they contradict commonly known facts.

Statement I : Pictures can tell a story.
Statement II: All storybooks have pictures.
Statement III: Some storybooks have words.

Conclusion: Some storybooks have both words and pictures.

✓ Correct Ans **A.** Conclusion is always true.
B. Conclusion is always false.
C. Conclusion is sometimes true.
D. Can't decide.

Explanation

Construct a Venn digram to arrive at the answer,

Q: 2 Gloriosa superba, commonly known as the flame lily, is species of flowering plant. It is a climbing lily species that is equally beautiful and poisonous. Every part of the plant is poisonous and it's toxic enough to cause human and animal fatalities if ingested. It has been used to commit murder, to achieve suicide, and to kill animals. This flower has **very unique** flame-like six tepals generally bright red to orange, with yellowish bases.

Niko: "A person has died after eating a random flower in a forest as the flower was looking beautiful. That flower must be a flame lily."

Soya: "Yesterday I saw there is a plant with flowers which has flame-like six tepals generally bright red to orange, with yellowish bases, so that must be a Gloriosa superba plant"

If the given information is true, whose reasoning must be correct?

A. Niko only

✓ Correct Ans **B.** Soya only

C. Both Niko nor Soya

D. Neither Niko nor Soya

Explanation

Gloriosa superba is beautiful and poisonous and it is poisonous enough to kill a human. But that doesn't mean that it is the only beautiful flower which is poisonous enough to kill a human. Therefore Niko is wrong.

As flame-like six tepals generally bright red to orange, with yellowish bases is unique to Gloriosa superba Soya is correct.

Q: 3 Robin loves dogs and rabbits but she hates cats. James loves pets and he has two dogs, a cat, and a rabbit. He loves them all. Ron doesn't like dogs, but he has a cat and two rabbits.
Which of the following is correct about the above situation?

A. Robin hates all the pets of James

B. Robin loves all the pets of Ron

C. Ron hates all the pets of Robin

✓ Correct Ans **D.** James loves all of Robin's pets

Explanation

Here are the pets that each person loves

- Robin – Dogs, Rabbits
- James – Dogs, Cats, Rabbits
- Ron – Cats, Rabbits

So, James loves all the pets of Robin.

Q: 4 You are given an A4 paper. If you fold it as shown in the dotted line in figure 1 and then cut along the dotted line shown in figure 2, which are the resulting shapes?

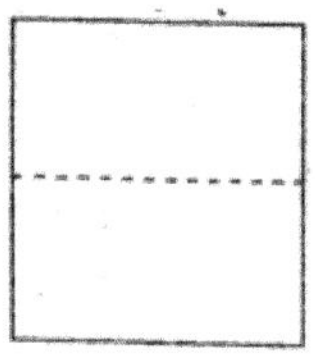

Figure 1

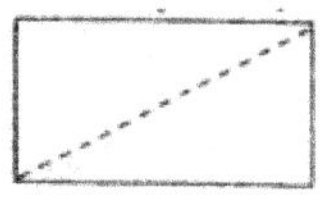

Figure 2

A. Four triangles

B. Two triangles

✓ Correct Ans **C.** Three triangles

D. A rectangle and two triangles

Explanation

You will get two smaller triangles from the below part from the cut.
From the above part, you will get a larger triangle.
Take a piece of paper and try.

Q: 5 **Neville**: "If you win the chess championship, you can beat Ron."

Jonathan: "I won a chess match with Ron. So, I can win the championship."

If Neville is correct, Jonathan is ...

A. Correct

B. Wrong

C. Both correct and wrong

✓ Correct Ans **D.** Cannot decide

Explanation

What Neville says is, that if Jonathan wins the championship, then he can defeat Ron. But that doesn't mean if Jonathan defeats Ron, he will win the championship.

Q: 6 A: "There is at least one student who got full marks for the assessment. "

B: "Everyone got at least 1 mark."

Which of the following is correct?

A. A and B have the same meaning

B. A has the opposite meaning of B

C. B is the opposite of A

✓ Correct Ans **D.** None of the above

Explanation

A → some students got full marks
Negation of A → no one got full marks
B → no one got 0 marks
Negation of B → some students got 0 marks
So, it's clear that there is no relation between A and B.

Q: 7 **If someone is good at running, he is also good at long jump.**
Athletes who are good at long jump are tall.

Which of the following is a possible conclusion based on above two statements ?

A. Tall people are good at running

B. Tall people are good at the long jump

C. Good runners are tall

✓ Correct Ans **D.** None of the above

Explanation

The statements can not be related. There may be athletes who are good at long jump, but not good at running.

Q: 8 In the school music band none of the member's height is less than 168cm.
All the members of the school band must know how to play the trumpet properly.

Darren: "Susan plays the trumpet very well also he has a height of 175cm. He must be a member in the school music band"

Noah: "Paul is willing to join the school music band but he is only 164cm. Definitely, he can't join the band"

If the information in the box is true, whose reasoning must be correct?

A. Darren only

✓ Correct Ans **B.** Noah only

C. Both Darren and Noah

D. Neither Darren nor Noah

Explanation

There may be additional criteria for joining. However, lower than required height is disqualifying criteria.

Q: 9 *A microorganism is a living organism which is too small to be seen with the eye but visible under a microscope. Viruses are even smaller therefore they can only be seen through an electron microscope. The term microorganisms do not include viruses, which are generally classified as non-living as they have no energy metabolism.*

Paul: "If there is an organism that doesn't have an energy metabolism, then that must be a virus."
Kia: "As viruses are not microorganisms, in some situations they might be seen with the eye"

If the given information is true, whose reasoning must be true?

A. Paul only

B. Kia only

C. Both Paul and Kia

✓ Correct Ans **D.** Neither Paul nor Kia

Explanation

Viruses have no energy metabolism doesn't certain that any organism that doesn't have an energy metabolism must be a virus. Therefore Paul is wrong.
As described in the passage Viruses are even smaller to see through a microscope therefore they can only be seen through an electron microscope. Therefore viruses can be never seen with eye. So Kia is also wrong.

Q: 10 *If Mr. Lucas gets the majority of the votes in the election, he will become a parliament member.*
If he becames a parliament member he will definitely speak for the low-income people who suffer from taxes, otherwise he will not speak.
If he speaks for the low-income people who are suffering from taxes the government might consider cutting some taxes.

If the given information is true, which one of the following must be true?

A. Mr. Lucas got the majority of the votes in the election but he didn't speak for the low-income people who suffer from taxes

B. Mr. Lucas did not get the majority of the votes in the election but he spoke for the low-income people who suffer from taxes.

✓ Correct Ans **C.** Mr. Lucas got the majority of the votes in the election and he spoke for the low-income people who suffer from taxes.

D. Mr. Lucas got the majority of the votes in the election and the government cut some taxes.

Explanation

If Mr. Lucas got the majority of the votes in the election he will become a parliament member and will speak for the low-income people who suffer from taxes. Therefore option C is correct.

Q: 11 *Lithium is a soft, silvery-white, alkali metal. Lithium metal reacts vigorously with water. Of the alkali metals,* ***only lithium reacts with nitrogen****, and it forms a nitride.*

Charles: "If an alkali metal is soft, silvery-white and reacts vigorously with water, then that must be Lithium"

Karen: "If there is an alkali metal that reacts with nitrogen and forms its nitride that must be Lithium"

If the information is true, whose reasoning must be correct?

A. Charles only

✓ Correct Ans **B.** Karen only

C. Both Charles and Karen

D. Neither Charles nor Karen

Explanation

Lithium is a soft, silvery-white, alkali metal and it reacts vigorously with water doesn't mean any metal with those qualities must be Lithium there might be many other metals with the same qualities too. So Charles is wrong.

As lithium is the only alkali metal that reacts with nitrogen, and it forms a nitride, Karen's reasoning is correct.

Q: 12 *Below is an excerpt from an article.*

Advances in technology have contributed to the meticulous use of robotics. The most important benefit of Robots is they are used to do repetitive actions or jobs considered too dangerous for humans. Robots are ideal for doing very risky jobs like nuclear cleaning, firefighting, searching and rescuing in disasters etc.

Fredric: "I don't agree with what the article prioritises. Robotic surgeries, allows doctors to perform many types of complex procedures with more precision, flexibility and control than is possible with conventional techniques. Robotic surgeries has reduced risk of infection or complications. There are a lot of benefits in a robotic surgeries than a regular surgery like less scarring because of smaller incisions and fewer sutures, briefer hospital stays and faster recovery, more rapid return to normal activities."

Which one of the following statements, if true, describes why Fredric disagrees with the article?

A. Robotics has been popular since before the development of technology.

B. Robotic surgery doesn't offer patient benefits when compared to traditional open surgery.

C. Robots cannot be used to do repetitive actions or jobs considered too dangerous for humans.

✓ Correct Ans **D.** Robotics has its most valuable achievements in the surgical field.

Explanation

Fredric tells that he disagrees with what the article prioritises. The article prioritises that using robots in dangerous life risking activities is the most important benefit of robotics. In Fredric's statement he is trying to convince that Robotic surgeries are the most important benefit of robotics. Therefore option C describes the best why Fredric disagrees with the article.

Q: 13 Following table contains the birthdays of 4 students.

Name	Birthday
Kevin	1998.05.20
Ross	1999.02.25
Helen	1999.01.20
Lily	1998.05.21

Who is the youngest?

A. Kevin

B. Helen

C. Lily

✓ Correct Ans **D.** Ross

Explanation

To be youngest, the birthday should be the latest. Latest year is 1999 and latest month in 1999 is February. So, Ross is the youngest.

Q: 14 **Sam** : "All the architects are good at drawing arts."
Ken : "Lily got full marks for art."

Which of the below is true according to **both Sam and Ken** ?

A. Lily is an architect

B. Lily can be an architect

C. Lily is going to be an architect one day

✓ Correct Ans **D.** No conclusion

Explanation

If someone is an architect, he is good at art drawing. But that doesn't mean that everyone who is good at arts is an architect or going to be an architect. So, there is no conclusion.

Q: 15 Julie, Michael and Rachel are discussing about current affairs.

Julie: "Every businessman owns more than one vehicle."
Michael: "If someone has many vehicles the maintenance cost is higher."
Rachel: "I do know that Ross is rich."

If all three are correct which of the following is a possible conclusion **about Ross** ?

A. Ross has one vehicle

B. Ross pays a lot for vehicle maintenance

C. Ross is a good friend of Rachel

✓ Correct Ans **D.** None of the above

Explanation

From the first two, if someone is a businessman, his vehicle maintenance cost is higher.
However, it is not known if Ross is a businessman. So, no conclusion can be arrived.

Q: 16 Cantharidin are toxic agent released by blister beetles which can cause irritation or blistering when it comes into contact with the skin, mouth, eyes, throat or even in the digestive tract when eaten. Cantharidin poisoning is most common in horses who consume contaminated alfalfa hay. Horses can die as a result of exposure. Cantharidin poisoning affects other species as well, including dogs, cats, rabbits, and humans.

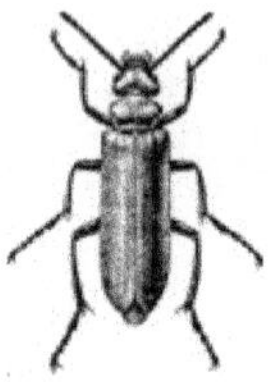

Which of the following must also be true?

A. Skin irritation can not occur when cantharidin came into contact with the skin.

B. Alfalfa hay with blister beetle should be fed to horses.

✓ Correct Ans **C.** Cats who eat a blister beetle can also suffer from cantharidin poisoning and it can be fatal.

D. Alfalfa hay which is suspected to not be contaminated must not be fed to horses.

Explanation

Based on the given information, a blister beetle has cantharidin which is a toxic agent that when released can cause poisoning or even death. Cats who eat a blister beetle can also suffer from cantharidin poisoning and it can be fatal which is option C.

Q: 17 Rose either drink coffee or cold water every day without fail. If it is cold and raining then Rose drinks coffee. If it is sunny and not raining then Rose drinks cold water. Sometimes it can be sunny and cold.

Which of the following statement must be true?

A. If it is raining and not sunny then Rose drinks her coffee.

✓ Correct Ans **B.** If it is raining and Rose does not drink her coffee then it is not cold.

C. If it is not sunny and it is cold then Rose drinks her coffee.

D. If it is sunny and not raining then Rose drinks her coffee.

Explanation

Option B must be true because it is stated that Rose drinks her coffee when it is cold and raining, without fail.

Q: 18 A terpene called camphor is frequently found in creams, ointments, and lotions. The oil produced by steam distillation after the camphor tree's wood is removed is known as camphor oil. To ease pain, irritability, and itching, use camphor topically. Camphor can induce convulsions, difficulty breathing, and even death when taken orally. High quantities of camphor can also be dangerous when applied topically or breathed. They may result in skin rashes or convulsions.

If the information is true, which of the following is true?

A. Camphor cannot cause skin irritation and must not be used.

B. Camphor can be mixed with a hot water and taken orally to relieve sore throat.

✓ Correct Ans **C.** Camphor must be used with caution and in small appropriate amounts.

D. Camphor can be mixed with tea and coffee to relieve stomach pain.

Explanation

Camphor can induce convulsions, difficulty breathing and even death when taken orally. It is also not said that camphor can be mixed with water, tea or coffee to relieve sore throat and stomach pain. Thus, option C is true. Camphor must be used with caution and in small appropriate amounts.

Q: 19 In the era of digital art tools, traditional painting has lost its significance.

Which of these statements weakens the above claim?

✓ Correct Ans **A.** Not all digital art tools offer the same level of creative freedom and depth as traditional painting.

B. Only professional artists can truly appreciate the value of traditional painting techniques.

C. Traditional painting has always been limited in terms of colour choices and texture effects.

D. Digital art tools have completely replaced traditional painting methods in modern art schools.

Explanation

Option A weakens the claim by suggesting that some digital art tools may not match the creative capabilities of traditional painting. It implies that traditional painting still offers unique advantages that are not fully replicated by digital tools.

Q: 20 Company A is calculating its annual sales. Which statements will help solve this calculation?

i. **Company A sells a piece of tile at $20, and it sold 10,000 pieces that year.**
ii. **The company only sells a single product.**

A. Statement i is enough

B. Statement ii is enough

✓ Correct Ans **C.** Both statements are necessary

D. Both statements aren't enough to calculate the sales

Explanation

Statement i will help to calculate the total sales of the tiles sold. Statement ii clarifies that it only sells a single product, which is tiles; hence, no more calculations are needed. Therefore, both statements are necessary.

Q: 21 A, B, C, D, E, F and G are in a straight line but not necessarily in that order.

i. A is next to G
ii. G is fourth from C's right
iii. D is between B and F
iv. C is 3rd from B's left and also on one end

Who are the people adjacent to A?

A. B and C

B. C and G

✓ Correct Ans **C.** G only

D. F only

Explanation

From the first statement, we have A -> G. 2nd statement, C -> G; hence, the arrangement is C->G->A. 3rd statement. B->D->F. 4th statement, B->C. So. We have B->D->F->C->G->A.

Q: 22 For one to be the chair of a committee of 15 members, there is a need for the maximum number of votes. If 5 people are contesting for the post, which of the following reasoning is correct?

Elizabeth: If every candidate gets at least one vote, the maximum the winner can get is 11 votes.
Emily: A win is guaranteed if a candidate gets 8 votes.

A. Elizabeth
B. Emily
C. None of them
✓ Correct Ans **D.** Both statements

Explanation

If each candidate gets a vote, the contended votes remain 10. So, a candidate can get all the 10 but the other 1, thus gathering 11 votes. Anything above that is impossible.
As for Emily, if a candidate gets 8, no one can score more than that since 15 – 8 is 7, less than 8.
So, both statements are correct,

Q: 23 A certain country gives various currency denominations different names. The names are alpha, beta, gamma, ray, and omega. If 5 alphas make a beta, 3 betas make a gamma, 4 gammas make a ray, and 5 rays make an omega, how many betas are in 2 omegas?

A. 60
B. 40
✓ Correct Ans **C.** 120
D. 150

Explanation

So, if 1 omega equals 5 rays, we have 5 x 2 = 10 rays. If each ray is equivalent to 4 gammas, the total gammas are 4 x 10 = 40 gammas. Lastly, each gamma has 3 betas; the number of betas is 3 x 40 = 120.

Q: 24 **According to certain research, tall men have quite large feet.**

If the statement is true, which of the following statements is true?

i. If your grandfather is taller than your father and wears a shoe size 10, then size 9 could be ideal for your father.
ii. If your grandfather and father are the same height, their shoe size can be the same.
iii. Whereas your father is shorter than your grandfather, his shoe size can be bigger than your grandfather's.

A. All the above
B. None of the above
✓ Correct Ans **C.** i and ii
D. i and iii

Explanation

Statements i and ii are correct since an individual's height and shoe size are directly proportional. The iii statement contradicts that finding.

Q: 25 ***Both Eva and Adam are good at Math. Only Eva is good at History. Adam is good at Science, which Eva isn't.***

Based on the above information about Eva and Adam, which one of the following must be correct?

A. Both of them are good at Math and Science.

✓ Correct Ans **B.** Only Adam is not good at History

C. Eva is not good at anything that Adam good at

D. Both of them are good at all the subjects

Explanation

From the the given information, out of Eva and Adam, Only Eva is good at History. So only Adam is not good at History

Q: 26 **If a student's attendance is more than 75%the teacher will add additional 5 marks for the written test, but only if the student has scored lower than 65 marks for the written test. To pass the written test a student must score at least 35 marks.**

Diana: "I already have 80% attendance. So I will definitely pass the written test even if I score 30 marks in my written test."

Samantha: "I also have 79% attendance. So if I get 95 marks in my written test, I will get 100 marks as my final results."

If the above information is true, whose reasoning must be true?

✓ Correct Ans **A.** Diana only

B. Samantha only

C. Both Diana and Samantha

D. Neither Diana and Samantha

Explanation

Additional 5 marks will be added to the final results of the exam of the students only who got both more than 75% attendance and scored less than 65 marks for the written paper. So Diana is correct. but Samantha is wrong .

Q: 27 **Analia has planned to go to a carnival and to take dinner at a nearby hotel.**

If Analia goes to the carnival she will likely go for a ride on the roller coaster.
If she goes on a roller coaster ride she will feel nauseous and dizzy.
If she feels nauseous and dizzy she will go back home without enjoying her dinner.

If the information given above is true, which of the following is not possible?

A. Analia didn't go on a roller coaster but she was nauseous and dizzy.

✓ Correct Ans **B.** Analia went on a roller coaster ride but she went to the hotel for dinner.

C. Analia went back home without enjoying her dinner but she even did not go on the roller coaster.

D. Analia went to the carnival but she did not go for a ride on the roller coaster.

Explanation

If Analia went on a roller coaster ride she would definitely feel nauseous and dizzy.
If she felt nauseous and dizzy she would definitely go back home without enjoying her dinner.

Therefore **option B** is not possible.

Q: 28 Rosy and Julie attended a life lesson workshop. The instructor told : "The best way to escape from a problem is to solve it." Instructor asked everyone about thier understanding on her statement.

Rosy: Your life will be dull if you don't face a problem.
Julie: To escape from problems, you should always have some solutions with you.

Who's explanation is more accurate to the meaning of instructor's statement ?

A. Rosy is accurateJulie is accurate

✓ Correct Ans **B.** Julie is accurate

C. Both Julie and Rosy are accurate

D. Neither Julie nor Rosy is accurate

Explanation

The instructor was providing problem solution strategy. There was no context of life without problems.
Julie reconfirmed her understanding that solution is only way to get rid of problems. So Julie's understanding is more accurate.

Q: 29 There are 5 trees in a garden. All the trees were planted for over the past 5 years. The 1st tree was planted in the 1st year and 2nd tree was in the second year and so on. The total heights of all the trees is 100 metres.

Madison: " All the trees may have different heights".
Addison: "All the trees will have same height."

Whose argument is correct?

✓ Correct Ans **A.** Madison is correct

B. Addison is correct

C. Both are correct

D. Both are wrong

Explanation

It is expected that older trees would have more height. The use of word 'may' makes Madison's argument correct.

Q: 30 **Statement I:** James is a good director.
Statement II: All directors are intelligent.

Based on above two statements, which conclusion is true ?

A. All intelligent are directors

✓ Correct Ans **B.** James is intelligent

C. Both of the above

D. None of the above.

Explanation

All directors are intelligent. James being a director (good or no so is not important here) is considered intelligent.

Q: 31 Maya's college is giving scholarships to students for further studies in France.

If a student has at least some knowledge of French, then they only have to pass the aptitude test. If they have none, then they either needed to get an excellent score on the aptitude test or follow the French course that the college offers and pass the French test and then pass the aptitude test.

Maya is fluent in French but failed to get the scholarship. What must have been the reason?

A. She didn't pass the French test.

B. She didn't get an excellent score on the aptitude test.

C. She failed the both French test and aptitude test

✓ Correct Ans **D.** She failed the aptitude test

Explanation

Since Maya is fluent in French, she only has to pass the aptitude test to get the scholarship.
Therefore she must have failed the aptitude test. The correct answer is D.

Q: 32 *If someone is a lawyer, then that person has good knowledge about the law and is good at arguing.*

Which of the following is the opposite of the above statement?

A. Some lawyersare good at arguing.

B. Some lawyers don't have good knowledge about the law.

✓ Correct Ans **C.** Some lawyers are not good at arguing or don't have good knowledge about the law.

D. Some lawyers are bad at arguing and have good knowledgeof the law.

Explanation

The negation of the given statement has the opposite meaning. That is, there are lawyers without good law knowledge or not good at arguing.

Q: 33 **In a survey of pets that elementary school students 6 were having dogs, 7 were having cats and 4 were having birds. Students had no other animals as pets.**

James: 17 students must have participated in the survey.

If the information is true, which one of the following sentences shows the mistake James has made?

A. Not only dogs, cats and birds there are so many other animals that kept as pets

✓ Correct Ans **B.** The number of those who participated in the survey might be less than 17

C. There must be more than 17 students who participated in the survey

D. The number of those who participated in the survey might be more than 17

Explanation

The same student can have more than one pet. So the total number of students who have each pet might not be the same number of students who participated. So the number of students who participated in the survey might be 17 or less than 17

Q: 34 *If lightning strikes today people won't go outside and tomorrow trees will look so green.*
It's not lightning today.

Which of the following is a possible conclusion?

A. People won't go outside today.

B. Trees will look so green tomorrow.

C. People won't go outside today, and trees will look so green tomorrow.

✓ Correct Ans **D.** None of the above.

Explanation

Given that, if it is lightning, then people won't go outside and trees will look so green tomorrow. But from that, we cannot conclude anything related to the case where it isn't lightening. So, we cannot conclude anything.

Q: 35 *If Rachel is wearing new clothes, then she has bought some new clothes or she has borrowed some clothes from Monica.*

Which of the following has the same meaning as the above statement?

A. If Rachel is wearing new clothes and she didn't borrow clothes from Monica.

B. If Rachel is wearing new clothes and she didn't buynew clothes.

✓ Correct Ans **C.** If Rachel didn't borrow clothes from Monica and didn't buy some new clothes then she is not wearing new clothes.

D. If Rachel is not wearing new clothes, then she didn't borrow clothes from Monica.

Explanation

The contrapositive of the given statement has the same meaning. That is if Rachel has not bought new clothes, she didn't borrow any clothes from Monica, then she is not wearing new clothes.

Q: 36 *Every bike is fast.*
Jonathan bought an expensive bike.

So, we can conclude that ..

✓ Correct Ans **A.** Jonathan's bike is fast.

B. Jonathan's bike isn't fast because it's expensive.

C. None of the above.

D. Cannot decide.

Explanation

From the first statement, if something is a bike, then it is fast. Jonathan bought a bike. So, it is obvious that his bike is also fast.

Q: 37 There is a 10-day music festival that will be held from 1st to 10th of this month. Ticket prices for the music festival are shown below.

- 1 day = $20
- 2 consecutive days = $30
- 5 consecutive days = $60

Paul wants to go to the music festival on the 1st, 2nd, 7th, 8th, 9th, and 10th. What is the minimum amount Paul can spend to go to the music festival on all these days?

A. $100

✓ Correct Ans **B.** $90

C. $110

D. $70

Explanation

Paul has several options.

1. He can buy 6 tickets for each day -> $120
2. He can buy one ticket for the first two days and then one per day -> $30 + $80 = $110
3. He can buy one ticket for the first two days and another ticket for the next 4 days -> $30 + $60 = $90

So, the least amount is $90.

Q: 38 **Professor Lucas:** "End semester exam follows mid semester exam. All the students who got more than 75 marks in the mid-semester exam will automatically get an A pass for the end semester exam, without taking end semester exam."

Helen: "Adam got 75 marks in the mid-semester exam. So, he will get an A pass for the end semester exam."

Kathy: "Chandler got 60 marks for the mid-semester exam. He may or may not get an A pass for the end semester exam."

Whose reasoning is correct according to Professor Lucas?

A. Only Helen

✓ Correct Ans **B.** Only Kathy

C. Both of them

D. No one

Explanation

According to Professor Lucas, no one can surely tell that Adam is going to get an A pass for the end semester exam, because he didn't get more than 75 marks for the mid-semester exam. So, Helen is wrong. However, Professor Lucas only mentioned a sufficient condition to get an A pass in the end semester exam. Anyone may be able to get an A pass for the end semester exam even without getting more than 75 marks for the mid-semester exam. So, only Kathy is correct.

Q: 39 Observe the following square braces containing four numbers.

[4, 6, 8, 32]

[3, 5, 7, 22]

[2, 4, 6, 14]

[5, 7, 9, ?]

If all four numbers in each brace follow the same rule, which number can replace the question mark?

A. 11

✓ Correct Ans **B.** 44

C. 108

D. 54

Explanation

The 4th number = 1st number x 2nd number + 3rd number

So, in the last square bracket, ? = 5 x 7 + 9 = 44.

Q: 40 *If Adam got sick, he won't be able to submit the assignment before the deadline.*

If Adam does not submit the assignment before the deadline, he cannot get an A+.

Adam didn't get an A+.

That means ..

A. Adam got sick.

B. Adam didn't get sick.

C. Adam didn't get sick and he didn't study well.

✓ Correct Ans **D.** No conclusion.

Explanation

If Adam got sick then he cannot get an A+. But this statement doesn't always work in the other direction. That means, if Adam didn't get an A+, we cannot conclude that he got sick. So, no conclusion can be made.

Answers And Solutions
Learnikx Education

Q: 1 According to Extract A, what is one benefit of good quality sleep ?

✓ Correct Ans **A.** Improved physical appearance
B. Increased stress levels
C. Reduced cognitive function
D. Impaired productivity

Explanation

The article mentions that quality sleep plays a vital role in maintaining a healthy complexion, reducing the appearance of dark circles, and puffiness, and promoting a more radiant and youthful appearance.

Q: 2 According to Extract A, what role does good quality sleep play in emotional well-being?

A. It increases susceptibility to negative emotions
✓ Correct Ans **B.** It improves emotional resilience
C. It exacerbates mood disorders
D. It causes chronic stress

Explanation

The article highlights that quality sleep promotes emotional resilience and stability, allowing individuals to better manage stress, regulate emotions, and improve overall mood.

Q: 3 According to Extract A, what is one aspect of physical health that is influenced by good quality sleep?

A. Increased risk of illnesses
B. Unhealthy weight gain
C. Weakened immune system
✓ Correct Ans **D.** Faster recovery from illnesses

Explanation

The article mentions that good quality sleep supports the immune system, reducing the risk of illnesses and promoting faster recovery. This indicates that sleep plays a role in the body's ability to recover from illnesses more efficiently.

Q: 4 According to Extract B, what are some physical effects of staying awake for 2 days?

A. Enhanced motor skills and coordination

B. Improved physical performance and strength

✓ Correct Ans **C.** Slower reflexes and reduced physical coordination

D. Increased energy levels and physical endurance

Explanation

The article highlights that lack of sleep can result in decreased motor skills, reduced strength, and slower reflexes, affecting physical coordination.

Q: 5 According to Extract B, how does sleep deprivation impact the immune system?

A. Boosts the immune response and strengthens immunity

B. Decreases the risk of infections and illnesses

✓ Correct Ans **C.** Weakens the immune system and increases susceptibility to illnesses

D. Does not affect the immune system

Explanation

The article mentions that sleep deprivation weakens the immune system, making individuals more susceptible to infections and illnesses.

Q: 6 According to the paragraph, how can individuals contribute to energy conservation in their daily lives?

A. Supporting green energy programs offered by utility providers.

B. Recycling and reusing products whenever possible.

C. Engaging in tree-planting initiatives.

✓ Correct Ans **D.** Turning off lights when not in use and using energy-efficient appliances.

Explanation

Individuals can contribute to energy conservation by adopting simple practices such as turning off lights when not in use and using energy-efficient appliances. These actions reduce energy consumption and help mitigate climate change, while also promoting cost savings and environmental sustainability.

Q: 7 Why is spreading awareness and educating others about global warming important?

A. It encourages others to engage in tree-planting initiatives.

B. It helps to increase individual energy consumption.

✓ Correct Ans **C.** It inspires collective action and promotes sustainable practices.

D. It supports the exploration of fossil fuel sources.

Explanation

Spreading awareness and educating others about global warming is important as it inspires collective action and encourages individuals to adopt sustainable practices. By increasing knowledge and understanding of the issue, more people can contribute to combating global warming and work towards creating a sustainable future.

Q: 8 How can individuals practice sustainable consumption to address global warming?

A. Choosing petrol powered vehicles and maintaining them properly.

✓ Correct Ans **B.** Participating in recycling initiatives and minimizing single-use plastics.

C. Supporting organizations dedicated to restoring forests.

D. Opting for green energy programs offered by utility providers.

Explanation

Individuals can practice sustainable consumption by participating in recycling initiatives and minimizing single-use plastics. By reducing waste and promoting recycling, individuals can help conserve resources and minimize the environmental impact of their consumption habits, thereby contributing to the fight against global warming.

Q: 9 What does the line "It is the blight man was born for" suggest about the nature of sorrow?

✓ Correct Ans **A.** Sorrow is natural and inevitable.

B. Sorrow can be avoided with effort.

C. Sorrow is temporary and fleeting.

D. Sorrow is caused by external factors.

Explanation

The line suggests that sorrow is natural and inevitable in human life.

Q: 10 What is the significance of the line "Sorrow's springs are the same"?

- **A.** It emphasizes the beauty of spring.
- ✓ Correct Ans **B.** It suggests that sorrow is eternal and unchanging.
- **C.** It describes the changing seasons.
- **D.** It highlights Margaret's love for nature.

Explanation

The line suggests that sorrow is eternal and unchanging, similar to the seasons.

Q: 11 What is the speaker's message to Margaret in the poem's closing lines?

- ✓ Correct Ans **A.** The speaker urges Margaret to embrace her grief.
- **B.** The speaker advises Margaret to forget her sorrows.
- **C.** The speaker reassures Margaret that her grief is temporary.
- **D.** None of the above.

Explanation

The speaker urges Margaret to embrace her grief and acknowledge its presence.

Q: 12 What does the poem imply about the relationship between age and the experience of sorrow?

- **A.** Sorrow becomes less significant with age.
- ✓ Correct Ans **B.** Sorrow deepens and becomes more understood with age.
- **C.** Sorrow has no meaning to children.
- **D.** Sorrow is incomprehensible to the elderly.

Explanation

The poem implies that as one grows older, they may have a deeper understanding of and experience with sorrow.

Q: 13 What is the overall tone of the poem "Stanzas" by Emily Bronte?

- **A.** Joyful and celebratory
- ✓ Correct Ans **B.** Melancholic and introspective
- **C.** Energetic and enthusiastic
- **D.** Suspenseful and mysterious

Explanation

The overall tone of the poem is melancholic and introspective, as the speaker reflects on their past and contemplates their choices.

Q: 14 Which poetic device is used in the line "And visions rising, legion after legion"?

A. Metaphor
B. Alliteration
C. Simile
✓ Correct Ans D. Personification

Explanation

The line uses personification, as it attributes human characteristics (visions rising) to non-human entities (visions).

Q: 15 In the poem, what do the "gray flocks in ferny glens" symbolize?

A. Wealth and prosperity
✓ Correct Ans B. Nature and simplicity
C. Loneliness and isolation
D. Historical figures and events

Explanation

The "gray flocks in ferny glens" symbolize the beauty and simplicity of nature, contrasting with the complexities of the world.

Q: 16 What does the speaker mean by "Its unsustaining vastness waxes drear"?

A. The vastness of the world is inspiring and uplifting.
✓ Correct Ans B. The vastness of the world is overwhelming and depressing.
C. The vastness of the world is filled with unreal dreams.
D. The vastness of the world is joyful and lively.

Explanation

The speaker finds the vastness of the world overwhelming and depressing, which is conveyed through the phrase "waxes drear."

Q: 17 What is the deeper meaning behind the line "More glory and more grief than I can tell"?

A. The speaker is indifferent to both glory and grief.
✓ Correct Ans B. Life contains both moments of great joy and deep sorrow.
C. The speaker is unable to experience either glory or grief.
D. Glory and grief are only illusions in life.

Explanation

The line suggests that life is a mixture of both moments of great joy (glory) and deep sorrow (grief), and the speaker acknowledges the complexity of human experience.

Q: 18 What does the speaker mean by "I'll walk where my own nature would be leading"?

✓ Correct Ans **A.** The speaker will follow their own instincts and desires.

B. The speaker will follow a predefined path chosen by others.

C. The speaker will walk in a place devoid of nature.

D. The speaker will walk only in heroic and historic places.

Explanation

The speaker expresses the intention to follow their own instincts and desires rather than conforming to societal expectations.

Q: 19 What does the speaker mean when they say, "The earth that wakes one human heart to feeling"?

✓ Correct Ans **A.** The earth has the power to make people feel emotions.

B. The earth has no effect on human emotions.

C. The earth can only be understood through logic and reason.

D. The earth is a source of great wisdom.

Explanation

The speaker suggests that the natural world has the power to evoke emotions and awaken human hearts to feeling, highlighting the connection between nature and human experience.

Q: 20 Which of the extracts refers to a place for performing arts?

A. Extract A

B. Extract B

C. Extract C

✓ Correct Ans **D.** Extract D

Explanation

According to extract D, the Opera House is a cultural gem that showcases the beauty of performing arts.

Q: 21 Which of the extracts refers to the tomb of pharaohs?

✓ Correct Ans **A.** Extract A

B. Extract B

C. Extract C

D. Extract D

Explanation

In extract A, the second line states that fact.

Q: 22 Which of the extracts mentions the place that provides a unique perspective of the city's landmarks?

A. Extract A

B. Extract B

✓ Correct Ans C. Extract C

D. Extract D

Explanation

In Extract C, it is mentioned that a ride on the London Eye provides a unique perspective of the city's landmarks.

Q: 23 Which of the extracts depicts the structure designed to protect from invasions?

A. Extract A

✓ Correct Ans B. Extract B

C. Extract C

D. Extract D

Explanation

Extract B mentions that great wall was built to protect the empire from invasions. It is a testament to the determination and ingenuity of the ancient Chinese people.

Q: 24 Which one fits in (1) ?

A. Sentence A

B. Sentence B

✓ Correct Ans C. Sentence G

D. Sentence H

Explanation

The sentence should be an introduction to the transformation to digital libraries.

Q: 25 Which one fits in (2) ?

✓ Correct Ans A. Sentence A

B. Sentence B

C. Sentence C

D. Sentence D

Explanation

The sentences should be on how the digital libraries are more easy than traditional libraries.

Q: 26 Which one fits in (3) ?

A. Sentence A

B. Sentence D

✓ Correct Ans **C.** Sentence B

D. Sentence E

Explanation

The paragraph says that the people can gain knowledge from anywhere. Thus, the corrects sentence must comply with the accessibility theme.

Q: 27 Which one fits in (4) ?

A. Sentence A

B. Sentence B

C. Sentence H

✓ Correct Ans **D.** Sentence E

Explanation

The prior sentences tell about the easy accessibility without having opening hours. Thus, the correct answer should tell about that.

Q: 28 Which one fits in (5) ?

A. Sentence A

B. Sentence B

✓ Correct Ans **C.** Sentence C

D. Sentence D

Explanation

The prior sentence says that digital libraries are known for cost saving. Thus, the correct sentence must prove that point.

Q: 29 Which one fits in (6) ?

A. Sentence A

B. Sentence B

C. Sentence C

✓ Correct Ans **D.** Sentence D

Explanation

The next parts of the sentence tell about the new ways people can learn by using digital libraries. The correct sentence also talks about a new learning experience.

Q: 30 Which one fits in (7) ?

A. Sentence A

B. Sentence G

✓ Correct Ans **C.** Sentence H

D. Sentence C

Explanation

The next part of the paragraph points out the advantages of the digital libraries. So, the correct answer should address that fact.

Answers And Solutions
Learnikx Education

Q: 1

Francisco needs to do quite a few errands.
He needs to stop by the dry cleaners, the post office, the hardware store, and the bank. There are a number of ways he can tackle his list of stops.

How many ways can he accomplish his tasks?

A. 10

B. 12

C. 16

✓ Correct Ans **D.** 24

Explanation

There are 4 stops; they can be arranged in factorial 4 ways.
Factorial 4 = 4 x 3 x 2 x 1 = 24

Q: 2

All the friends gathered at Teresa's home for a party. Mila came 5 minutes before the party starts. The party started at 20:00. Mila took 1 hr 5 minutes to drive from her home to Teresa's home.

When did she leave her home?

✓ Correct Ans **A.** 18:50

B. 18:55

C. 17:55

D. 17:50

Explanation

Mila arrived at 7:55. Go backwards 1 hr 5 mins from 7:55, to reach 6:50 i.e. 18:50 in 24 hr format.

Q: 3 Donald, Tom and Lily had a total of 640 arrows.
Donald had 68 more arrows than Tom.
Lily had twice as many arrows as Donald.

How many arrows did Tom have?

A. 108

✓ Correct Ans B. 109

C. 210

D. 354

Explanation

Tom had n.
Donald = n + 68
Lily = (n + 68) x 2 = n x 2 + 136

Total = n + n + 68 + nx2 + 136 = 640
n x 4 = 640 - 136 - 68 = 436
n = 436 / 4 = 109

Q: 4 A box measures 56 cm long, 23 cm wide and 106 cm high. Elise wants to fill the box with as many cubes as she can. Each cube has a volume of 64 cu. cm.
What is the maximum number of cubes that Elise can fill the box with?

A. 1110

B. 1860

✓ Correct Ans C. 1820

D. 2000

Explanation

Cube has length, width and height as 4 cm.

Heightwise we may put 106/4 = 26 cubes (discarding remainders).
Widthwise we may put 23/4 = 5 cubes
Lengthwise we may put 56/4 = 14 cubes

14 x 5 x 26 = 1820 ; Elise can fill the box 1820 cubes.

Q: 5 Jason finished reading all 185 pages of a story book in 5 days. For each day, he read 10 pages more than the previous day.
How many pages did he read on the first day?

✓ Correct Ans **A.** 17
B. 18
C. 23
D. 35

Explanation

First day = n
Second day = n + 10, Third day = n + 20, and so on.
n + n+10 + n+20 + n+30 + n+40 = 185
n = 17

Q: 6 Select the suitable letters for the empty boxes.

A	B	C	D
F	E	D	(i)
E	F	G	(ii)
J	I	H	G

A. E, H
B. F, H
✓ Correct Ans **C.** C, H
D. C, F

Explanation

- 1st row → ABCD (forward)
- 2nd row → CDEF (backward)
- 3rd row → EFGH (forward)
- 4th row → GHIJ (backward)

Q: 7 You can see 11 pillars of its fence when you are looking from each of the three sides of an equilateral triangular land. How many pillars are there in the fence?

A. 33
B. 27
✓ Correct Ans **C.** 30
D. 31

Explanation

Three pillars on vertices. Other than that, 9 pillars on each side. So, number of pillars = 9 + 9 + 9 + 3 = 30.

Q: 8 You need to make the following shape a square. How many more small squares will you need minimum ?

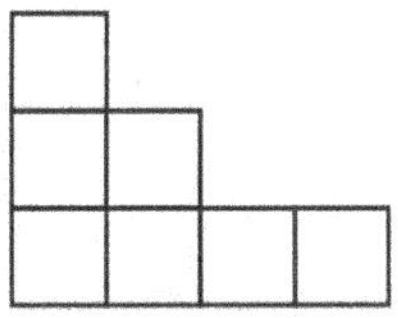

A. 6

B. 7

✓ Correct Ans **C.** 9

D. 10

Explanation

To make a square, there should be 4 small squares on each side. Small squares to be added :

- 1st row → 4
- 2nd row → 3
- 3rd row → 2
- 4th row → 0
- Total → 9

Q: 9 Jennie is in a bus queue. There are 4 people behind her and 6 people in front of her. If Jennie's brother was in the middle of the queue, which of the following is correct?

A. He is the 5th of the queue

✓ Correct Ans **B.** He is the 6th of the queue

C. He is the 7th of the queue

D. He is the 8th of the queue

Explanation

Number of people in the queue = 4 + 6 + 1 = 11. So, the middle position is the 6th.

Q: 10 David has to go 1400 m to school. Yesterday he left home for school and after riding his bicycle for 600 m, he remembered that he forgot the water bottle. So, he returned home and again left for school. How far has David gone?

A. 2000 m

✓ Correct Ans **B.** 2600 m

C. 3200 m

D. 1400 m

Explanation

Here are the distances :

- Went 600 m
- Returned 600 m
- Again went 1400 m

So, total = 600 + 600 + 1400 = 2600 m.

Q: 11 Siti wanted to buy 30 markers but she was short of $14. She bought 22 markers instead and had $18 left. How much money did Siti have?

A. $4

B. $32

✓ Correct Ans **C.** $106

D. $18

Explanation

8 markers → 14 + 18 = 32
1 marker → 32 ÷ 8 = $4
$4 X 22 + $18 = $106

Q: 12 Kavya bought 28 mangoes and pomelos. Each mango cost $3 and each pomelo cost $5. She paid a total of $118 for all the fruits.
How many pomelos did she buy?

✓ Correct Ans **A.** 17

B. 13

C. 32

D. 89

Explanation

Assume all are mangoes.
Cost → $3 X 28 = $84
Difference → $118 - $84 = $34
$5 - $3 = $2

No. of pomelos → $34 ÷ $2 = 17

Q: 13 The figure below is a net.

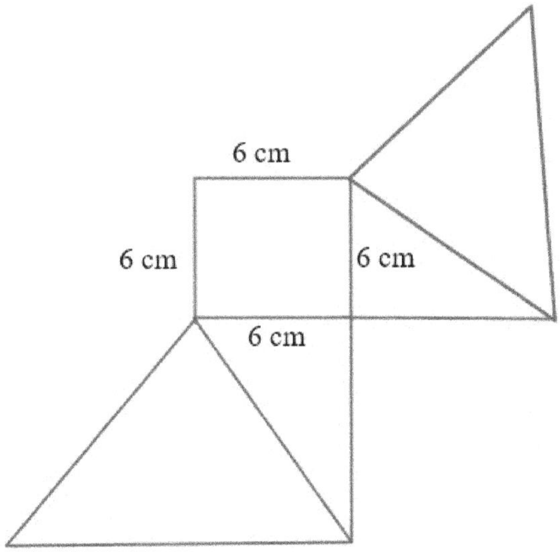

If the net is folded, what object is going to be formed?

A. Cuboid

B. Cylinder

✓ Correct Ans **C.** Pyramid

D. Cone

Explanation

Pyramids have triangular faces known as lateral faces. The lateral faces in the above-drawn shape are congruent isosceles triangles and the pyramid consists of these properties. Also, each edge of the base in a pyramid is usually connected to an apex. When the drawn shape is joined, the pyramid appears as shown below

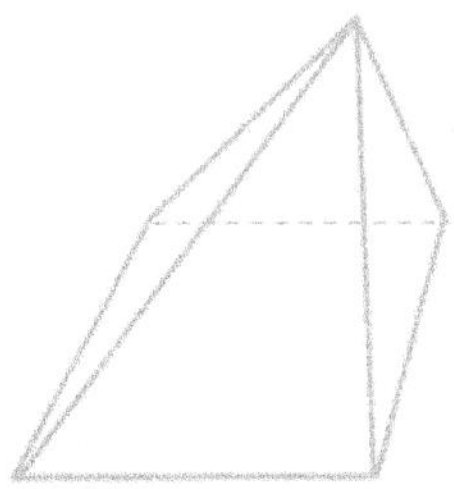

Q: 14 Four points P, Q, R, and S are such that PS is parallel to QR and PQ is parallel to RS. Show the possible positions of P, Q, R, and S.

✓ Correct Ans **A.**

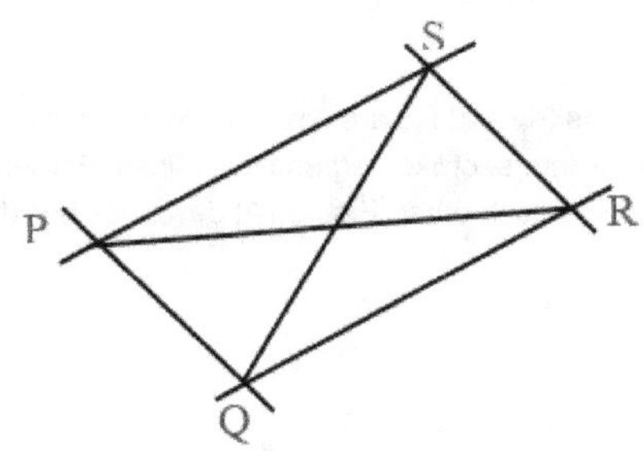

B.

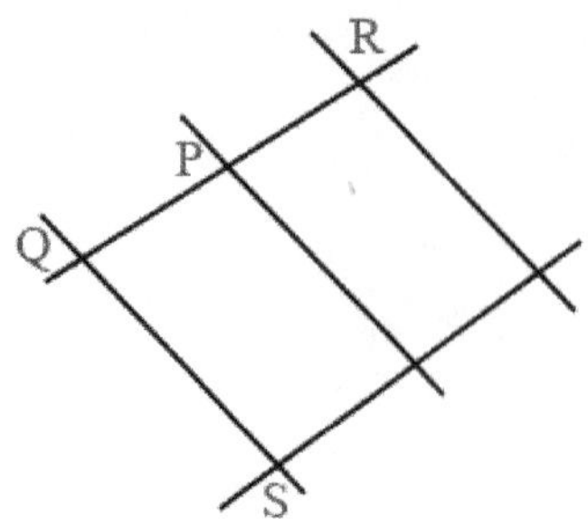

C.

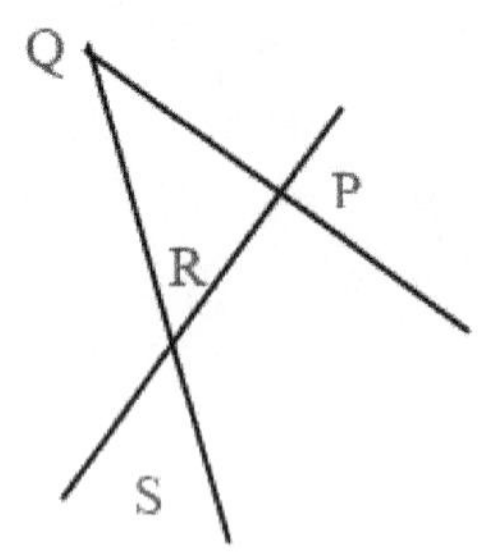

D.

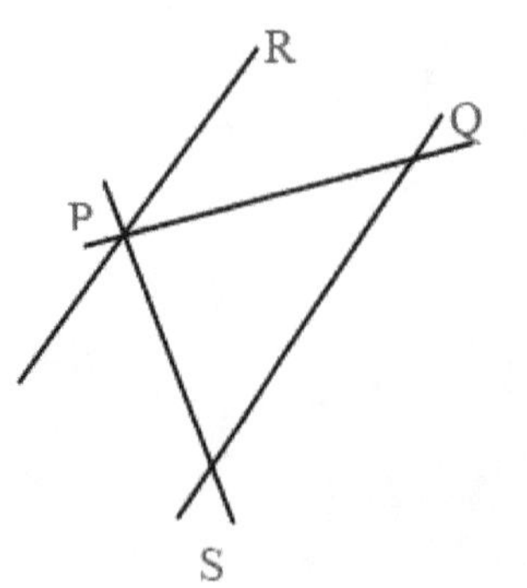

Explanation

The lines PS and QR are parallel to each other while lines PQ and RS are also parallel to each other. Lines PR and QS are perpendicular to each other. Parallel lines do not intersect like perpendicular lines. Perpendicular lines touch each other at a certain point but parallel lines do not cross each other. Figure (A) describes best these points' location.

Q: 15

Point M ($\frac{1}{2}$, 1) is the mid-point of points A (a, -3) and B (4, b). Find a and b.

A. -2, 1

✓ Correct Ans **B.** -3, 5

C. 3, -1

D. 5, 3

Explanation

Midpoint is the halfway between two endpoints. All you need to do is divide the sum of the x -values and the sum of the y -values by 2.

Midpoints of $x = \frac{a+4}{2} = \frac{1}{2}$.

When you multiply both sides by 2 you get a + 4 = 1. Hence, a = 1 – 4 = -3.

Midpoints of $y = \frac{-3+b}{2} = 1$.

When you multiply both sides by 2 you get -3 + b = 2.

Hence, b = 2 + 3 = 5.

Q: 16

The table below shows the average ages that people live in different countries. This is called 'life expectancy'. Work out the range for men and the median life expectancy for women.

Country	Life expectancy	
	Men	Women
UK	72	78
Britain	43	46
Australia	62	68
Germany	57	61
Poland	75	80
France	63	66

A. Men range = 57, Women median = 46

B. Men range = 43, Women median = 46

✓ Correct Ans **C.** Men range = 32, Women median = 67

D. Men range = 72, Women median = 68

Explanation

The biggest value and the smallest values for men's life expectancy are 75 and 43 respectively. Hence, the range for men is 75 – 43 = 32.

You need to put all the data values of women's life expectancy in order of size to get the median. Start from the smallest to the largest since the median is known as the middle value. The order of women's life expectancy values is 46, 61, 66, 68, 78, and 80.

If you get an even number of data values then there will not be one exactly in the middle. Work out the mean of the middle two numbers to find the median.

The mean is $\frac{66+68}{2} = \frac{134}{2} = 67$.

Therefore, the range of men and the median life expectancy for women is 32 and 67 respectively.

Q: 17 Jenny entered a diving competition. Each of the dives that Jenny makes has a dive rating. The harder the dive the higher the rating. Here are the rules for working out a score:

i. Remove the highest and the lowest mark
ii. Add together the middle three marks
iii. Multiply the answer of (ii) by the dive rating.

Each mark is out of 10. Jenny dives with a rating of 3.25. The judge gives the marks: **7.5, 7.7, 7.9, 7.3, 7.6**

Work out Jenny's score.

A. 80.2

B. 76.7

✓ Correct Ans **C.** 74.1

D. 53.4

Explanation

Arrange Jenny's marks from the lowest to the largest: 7.3, 7.5, 7.6, 7.7, 7.9
Follow rule 1 and remove the highest and the lowest mark to remain with 7.5, 7.6 and 7.7
Follow rule 2 and add together the remaining three marks as: 7.5 + 7.6 + 7.7 = 22.8
Now multiply the answer 22.8 by the dive rating which is 22.8 × 3.25 = 74.1

Q: 18 The diagram below represents two flashing lamps A and B. The lamps are switched on at the same time. Work out how often the two lamps flash together.

Lamp A flashes 20 seconds.

Lamp B flashes 35 seconds

A. 100 seconds

✓ Correct Ans **B.** 140 seconds

C. 200 seconds

D. 160 seconds

Explanation

The lights flash at:
A = 20, 40, 60, 80, 100, 120, **140**, 160......seconds
B = 35, 70, 105, **140**, 175......seconds
Using the LCM of 20 and 35, the flashlights flash together every 140 seconds.

Q: 19 Jason and Mark each think of a prime number that is greater than 100 and less than 110.
The total **sum of the digits** of Jason's number is 4 and the total sum of the digits of Mark's number is 10.

What number do you get when Jason's number and Mark's number are added together?

A. 203

✓ Correct Ans **B.** 212

C. 214

D. 109

Explanation

List all the prime numbers greater than 100 and less than 110. The prime numbers are 101, 103, 107, and 109.
Work out Jason's number whose values add up to 4. The number is 103 since the total sum of its values = 1 + 0 + 3 = 4.
Work out Mark's number whose values add up to 10. The number is 109 because the sum of its values = 1 + 0 + 9 = 10. Now add Jason's and Mark's numbers.
When you add Jason's and Mark's numbers you get 103 + 109 = 212.

Q: 20 Ravi bought 18 more egg tarts than Ayesha. Jessie bought 3 times as many egg tarts as Ravi. All the three girls bought a total of 187 egg tarts.
How many egg tarts did Ravi and Ayesha buy altogether?

A. 55

✓ Correct Ans **B.** 64

C. 60

D. 70

Explanation

Ravi = n, Ayesha= n-18, Jessie = nx3
Total = n + n-18 + nx3 = nx5 - 18 = 187
nx5 = 205 ; n = 41
Ravi + Ayesha = nx2 - 18 = 41x2 - 18 = 64

Q: 21 An eraser cost 30 cents and a packet of 5 erasers cost $1. Jonathan bought exactly 27 erasers.
What was the least amount of money that Jonathan spent on the erasers?

A. $5.50

B. $4.00

✓ Correct Ans **C.** $5.60

D. $6.00

Explanation

1 Eraser → $0.30
Packet of 5 Erasers → $1
Nearest multiple of 5 to 27 = 5 x 5 = 25 erasers
Number of erasers extra = 27 - 25 = 2 erasers
Least amount of money spent → (5 x $1) + (2 x $0.30) = $5.60

Q: 22 The mass of 2 identical packets of salt and 3 identical packets of sugar is 3862 g. The mass of 15 such packets of salt and 3 such packets of sugar is 8490 g.
What is the mass of one such packet of salt (in gms)?

✓ Correct Ans **A.** 356

B. 348

C. 376

D. 329

Explanation

2 Salt + 3 Sugar = 3862
15 Salt + 3 Sugar = 8490
Subtracting, 13 Salt = 8490 - 3862 = 4628
4628 ÷ 13 = 356g

Q: 23 The table below shows the rate of parking a car in a car park.

Parking Charges	
First Hour	$2.50
Every additional half hour or part thereof	$1.20

How much does Mr Bhim need to pay for parking his car from 7.30 a.m. to 9.15 a.m.?

A. $7.80

B. $6.50

C. $5.00

✓ Correct Ans **D.** $4.90

Explanation

First hour --- 7.30 a.m. to 8.30 a.m.
8.30 a.m. to 9.15 a.m. --- 45 minutes.
45 minutes consists of 1 half hour and another part thereof.
$1.20 + $1.20 + $2.50 = $4.90

Q: 24 A company ordered 25 potted plants to line up along their corridor from one end to the other end at equal spacing of 1.2 m apart. 4 of the potted plants broke during delivery and could not be used. The company used the remaining potted plants to line up along their corridor from one end to the other end at a new equal spacing. What was the new equal spacing between two potted plants?

A. 2 m

✓ Correct Ans **B.** 1.44 m

C. 1.64 m

D. 1.50 m

Explanation

25 potted plants = 24 intervals
Length of corridor = 24 x 1.2m = 28.8m
25 – 4 = 21
21 potted plants = 20 intervals
28.8 ÷ 20 = 1.44

Q: 25 The average height of a group of girls was 155 cm.
When 15 boys joined the group, the average height became 158 cm.
Given that the average height of the 15 boys was 165 cm, find the number of girls in the group.

A. 45

B. 25

✓ Correct Ans **C.** 35

D. 75

Explanation

There are n girls.
Total girls height = nx155

Total boys height = 15 x 165 = 2475

Total height = (n+15) x 158 = nx158 + 2370

nx155 + 2475 = nx158 + 2370;
2475 - 2370 = nx3; nx3 = 105; n = 35

Q: 26 The following pie graph shows the number of certain players of a particular country playing different games. Study the pie graph carefully and answer the following questions:

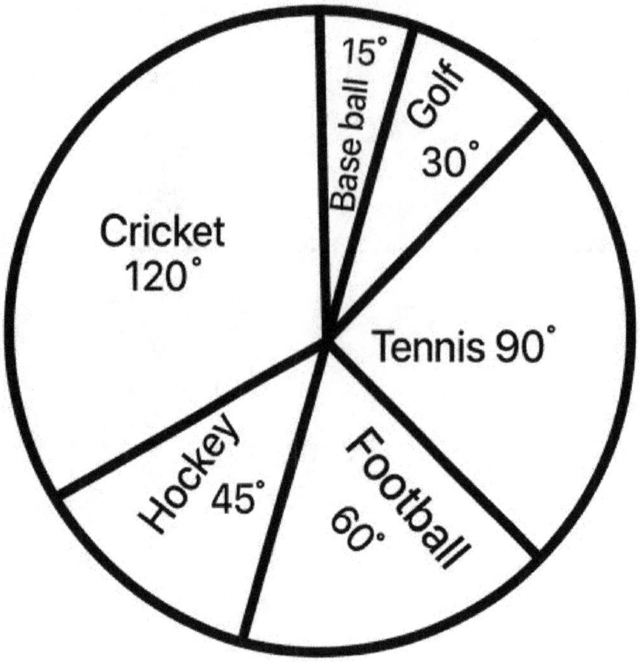

If the total number of persons is 90,000, then the difference between the number of persons playing only Tennis and Hockey is ...

✓ Correct Ans **A.** 11,250

B. 8,250

C. 7,500

D. 6,750

Explanation

360° = 90,000
Tennis = 90/360 = x/90,000 => 22,500
Hockey = 45/360 = x/90,000 => 11,250
22,500 – 11,250 = 11,250

Q: 27 A shipping company charges customers based on the weight of their packages. The weight **categories** and corresponding rates are as follows:

Upto 5 kg: $10
5.1 kg to 10 kg: $15
10.1 kg to 15 kg: $20
15.1 kg and above: $25

If a customer's package weighs 12.5 kg, how much will they be charged?

A. $10

B. $15

✓ Correct Ans **C.** $20

D. $25

Explanation

The weight of the package is 12.5 kg, which falls within the range of 10.1 kg to 15 kg. According to the rates, this weight category has a charge of $20.
Therefore, the customer will be charged $20 for their 12.5 kg package.
Among the given answer choices, option (C) $20 is the correct answer.

Q: 28

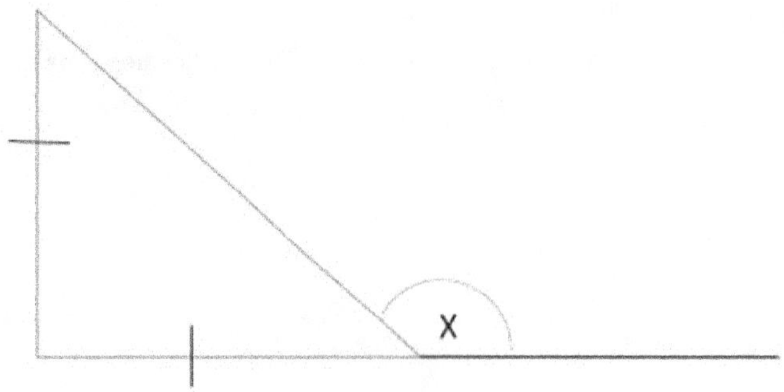

Given above is a right angled isosceles triangle. What is the value of x?

✓ Correct Ans **A.** 135 °

B. 125 °

C. 120 °

D. 115 °

Explanation

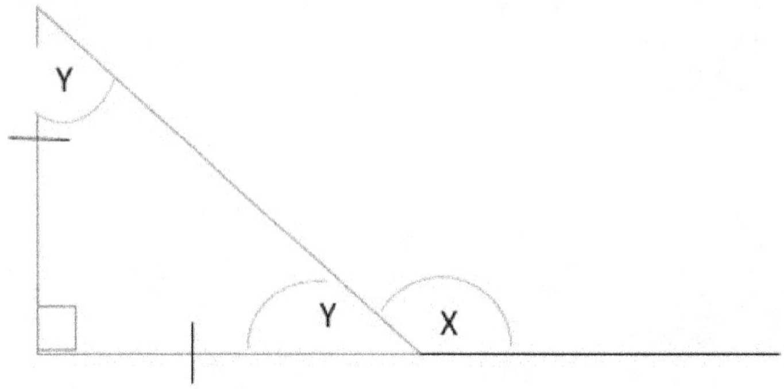

Since two sides are equal in length, the angles opposite to these sides are equal. So the remaining interior angles are equal.

2y= 180 °- 90 ° = 90 ° ; y = 45 °

BCD is a straight line. Sum of the angles on a straight line are equal to 180 °

So, 180 °- y = x ; 180 ° - 45 °=x ; x= 135 °

Q: 29

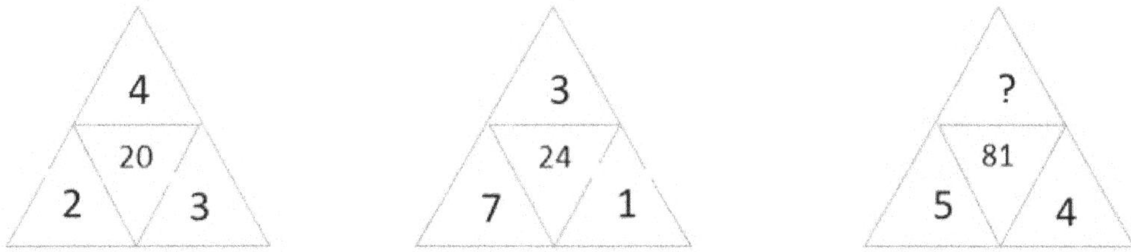

Identify the pattern and replace '?' mark.

A. 2

B. 6

✓ Correct Ans **C.** 9

D. 13

Explanation

Add bottom 2 numbers and multiply that answer from the top number to obtain the middle number.
(2+3) ×4 =20
(7+1) ×3 =24
(5+4) ×? =81
? = 81/9 =9

Q: 30 Handmade paper flowers are put in a bag. They are all identical in shape. There are 6 yellow flowers, 9 white flowers, 12 red flowers and 8 blue flowers.
What is the probability of taking out a flower, which is yellow, blue, white or red?

A. 0

B. $^{4}/_{35}$

C. $^{1}/_{35}$

✓ Correct Ans **D.** 1

Explanation

There are only yellow, blue, white and red flowers in the bag. If someone takes out a flower it must be yellow, blue, white or red. So the probability of taking out a flower that is yellow, blue, white or red is 1.

Q: 31 Below pie chart illustrates favorite sports of grade 7 students. Total no. of students is 150.

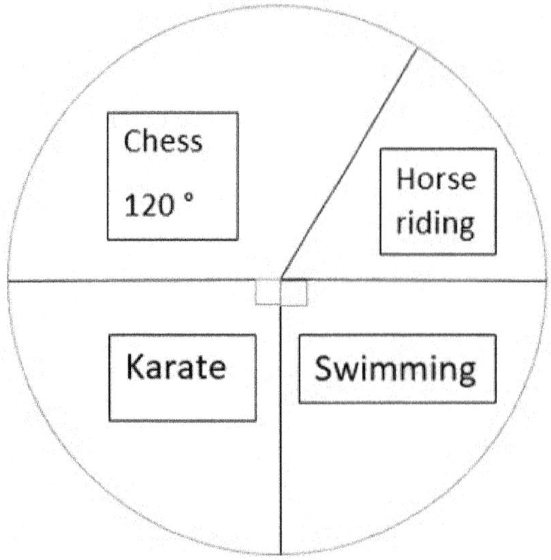

How many students like chess?

✓ Correct Ans **A.** 50

B. 60

C. 70

D. 80

Explanation

150 × 120° / 360° = 50 students like chess

Q: 32 In a house there is a huge window which has a height of 3m and width of 4m. Amelia decides to paste a painting on it leaving a border of 50cm wide all around it. What is the area of the border?

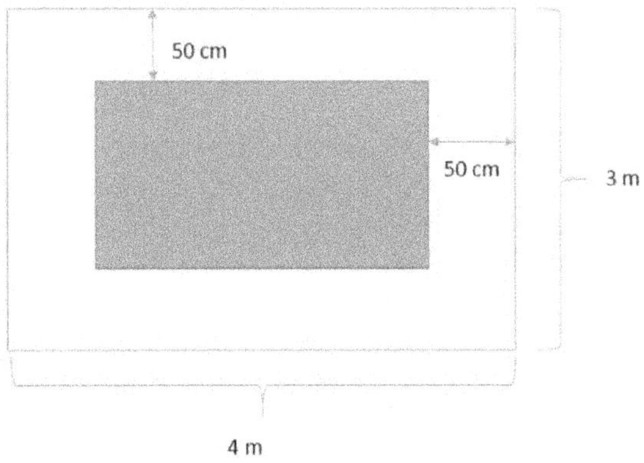

A. 12 m^2

B. 8 m^2

✓ Correct Ans **C.** 6 m^2

D. 4 m^2

Explanation

Area of the window = 4 ×3 = 12 m^2
Width of the painting = 400 cm – 50 -50 = 300 cm = 3m
Height of the painting = 300 cm - 50 -50 = 200 cm = 2m
Area of the painting = 3 ×2 = 6 m^2
Area of the border = 12- 6 = 6 m^2

Q: 33 Joe and Alan had a total of 33 marbles at first. After Joe lost 3 of his marbles and Alan lost half of his marbles, they had an equal number of marbles left.
How many marbles did Alan have in the end?

A. 14

✓ Correct Ans **B.** 10

C. 30

D. 12

Explanation

Start : Joe = 33 - n, Alan = n
End : Joe = 30 - n, Alan = $n/2$

$n/2$ = 30 - n ; n = 60 - 2n
3n = 60 ; n = 20

Alan at end = $n/2$ = 10

Q: 34 Carol, Dora, and Eve planted 11,753 roses altogether. Carol planted half as many roses as Dora and Eve planted half as many roses as Carol.
How many more roses did Dora plant than Eve?

A. 5073

✓ Correct Ans **B.** 5037

C. 5057

D. 5017

Explanation

Eve = n, Carol = 2n, Dora = 4n; Dora - Eve = 3n
Total = 7n = 11753; n = 1679
nx3 = 5037

Q: 35 Kelly took 4 hours to paint a room. Richard took 8 hours to paint an identical room. If both of them worked together, how long (**in hours**) would they take to paint the room?

✓ Correct Ans **A.** $2\ ^{2}/_{3}$ hrs

B. $2\ ^{1}/_{3}$ hrs

C. $4\ ^{1}/_{3}$ hrs

D. 2 hrs

Explanation

In 1 hour :
Kelly could paint $^{1}/_{4}$ the room.
Richard could paint 1/ 8the room.
$^{1}/_{4} + ^{1}/_{8} = ^{2}/_{8} + ^{1}/_{8} = ^{3}/_{8}$

Kelly and Richard could paint 3/8 of the room if they worked together.
$^{3}/_{8}$ of the room → 1 h
$^{1}/_{8}$ of the room → $^{1}/3$ h
$^{8}/_{8}$ of the room → $^{8}/_{3}$ h = $2\ ^{2}/_{3}$ h
They would need $2\ ^{2}/_{3}$ h to paint the room.

Q: 1

If Mary goes to the beach she will play with sand.
If she plays with sand she might collect sand into a bucket to take home.
If she has sea sand she will design wall art under the theme 'beach'.
Only if she designs satisfying wall art she will exhibit it in the art gallery.

If the given information is true, which one of the following is not correct?

✓ Correct Ans **A.** If Mary designs wall art under the theme 'beach' her art must be exhibited in the art gallery.
B. If Mary goes to the beach she might design a wall art under the theme beach.
C. If Mary's art was exhibited in the art gallery then she must have designed a satisfying wall art.
D. Even though Mary didn't go to the beach her art was exhibited in the art gallery.

Explanation

Only if Mary designs satisfying wall art she will exhibit it in the art gallery. Therefore if the wall art she made under the theme 'beach' is not satisfying to her she will not exhibit it in the art gallery. So option A is wrong.

Q: 2

Both Maya and Rima are gifted students. Maya is especially good at Math. Both of them are very good at science. Both of them got selected for the final rounds of Math and science quiz competitions representing their school, along with other students.

Students from their school won both Math and science best prizes.

If the above information is true, which of the following is correct?

A. If Maya is not the student who won the Math prize, definitely Rima is not the person who won that prize.
B. Maya must have won the Math prize and Rima have won the science prize.
C. Both Maya and Rima have won prizes.
✓ Correct Ans **D.** Neither Maya nor Rima might win any prize in quiz competitions.

Explanation

The information given doesn't tell Maya and Rima are the only students who participated in the competition representing their school. Therefore option D is correct.

Q: 3 *The vet clinic opens at morning 8 O' clock every weekend.*
Amalia and Robinson came to the clinic with their pets, but it was closed.

Amalia: "As today is Sunday and it's 8:20 a.m. the clinic should be open by now."

Robinson: "Today the veterinary surgeon must be on a holiday, let's go back."

If the information is true, whose expecation is **most likely** be correct?

✓ Correct Ans **A.** Amalia only

B. Robinson only

C. Both Amalia and Robinson

D. Neither Amalia nor Robinson

Explanation

As the vet clinic opens in the morning at 8 O'clock every weekend Amalia is correct. but being late to open the clinic doesn't mean the veterinary surgeon must be on a holiday, therefore Robinson is wrong.

Q: 4 **Every mathematician has a higher IQ.**
Neil is a mathematician.

So, we can conclude that ...

✓ Correct Ans **A.** Neil has a higher IQ.

B. Neil's IQ can not be measured.

C. Neil doesn't have a higher IQ.

D. No conclusion

Explanation

If someone is a mathematician, then he has a higher IQ. So, Neil has a higher IQ.

Q: 5 *Below are four statements. There may be additional statements that may not be required to arrive at a conclusion.*

"Shipowners are rich."
"Rich people have SUVs."
"Kevin has an SUV."
"New SUV models are mostly hybrid."

Which is the most correct conclusion?

A. Kevin is surely a shipowner

B. Kevin is surely not a shipowner

✓ Correct Ans **C.** Kevin might be a shipowner

D. Kevin is rich but not a shipowner

Explanation

From the first two statements, we can conclude that shipowners have SUVs. But being a shipowner is just a sufficient condition to own an SUV. There can be other people too, who own SUVs. But there is a possibility that Kevin is also a shipowner.

Q: 6 **An astronaut must have excellent observation skills and the ability to work under intense pressure.**

If the given information is correct which of the following must be true?

A. If Andrew has both excellent observation skills and the ability to work under intense pressure he must be an astronaut

✓ Correct Ans **B.** If Andrew has only the ability to work under intense pressure he cannot be an astronaut.

C. If Andrew has only the ability to work under intense pressure he might be an astronaut

D. If Andrew has excellent observation skills he must be an astronaut

Explanation

An astronaut must have excellent observation skills and the ability to work under intense pressure. Therefore if someone doesn't have either of these cannot be an astronaut. Option B is correct.

Q: 7 According to new studies, all octopuses and cuttlefish are venomous. But **Blue-ringed octopuses remain the only one dangerous to humans**. Blue-ringed octopuses can kill humans by biting and injecting venom.

If the information given is true, which of the following must be true?

A. If an animal died because of an octopus attack, that octopus must be a Blue-ringed octopus

B. None of the humans can be killed by an octopus attack

✓ Correct Ans **C.** If a human died as a result of an octopus attack that octopus must be a Blue-ringed octopus

D. According to new studies, people should stop eating cuttlefish as they are venomous.

Explanation

As Blue-ringed octopuses remain the only ones dangerous to humans and can kill humans by biting and injecting venom. Therefore if a human died as a result of an octopus attack that octopus must be a Blue-ringed octopus.

Q: 8 A description of the four students who scored the highest marks in each class is given below.

- *Sam scored twenty marks more than Kai*
- *Roy scored twenty-one marks less than Sam*
- *Henry scored two marks more than Roy*

Based on the above information, which one of the following must be true?

✓ Correct Ans **A.** Henry is the second.

B. Roy is the first.

C. Sam is the third.

D. Neither Sam nor Kai is the first.

Explanation

By analyzing the given information, Sam must be the first, Henry is the second, Kai is the third and Roy is the fourth in the section. Therefore A is correct.

Q: 9 **In Monica's company :**

Anyone who has a printer at their home has a laptop.
Anyone who has an electric car has a solar panel at their home.
Sam and Robin are also working in Monica's company.

Jack: "Sam has a solar panel in his house but he doesn't have an electric car."

Patricia: "Robin **used to have a printer** but he does not have a laptop"

If the given information is correct, whose reasoning is **impossible**?

A. Jack only

B. Patricia only

C. Both Jack and Patricia

✓ Correct Ans **D.** Neither Jack nor Patricia.

Explanation

Anyone who has an electric car has a solar panel at their home, doesn't mean there cannot be anyone who has a solar panel but he doesn't have an electric car. So Jack's statement is possible.
Anyone who has a printer at their home has a laptop. Robin used to have a printer, so now he may doesn't have a printer. Therefore Patricia's statement is also possible.

Q: 10 **Identify reasonable conclusion(s) following the statement.**

Statement: For all the TV sets manufactured in Australia, the 'Solar' brand has the largest sale.

Conclusion I: Volume of sales of all brands of TV sets manufactured is known.
Conclusion II: The production of no other TV set is as large as that of 'Solar'.

✓ Correct Ans **A.** Only conclusion I follows

B. Only conclusion II follows

C. Neither conclusion I nor conclusion II follows

D. Both the conclusion follows

Explanation

The statement has '**largest**' associated with **Sales**, not with Production. Hence conclusion II is incorrect.

Q: 11 Please refer to the statements below :

A: Kathy's cakes are tasty but expensive.
B: She gets fewer orders than other cake shops.

Which of the following shows the relation between A and B?

A. A is the effect and B is the cause

✓ Correct Ans **B.** A is the cause and B is the effect

C. A and B are effects of a different cause

D. A and B are independent

Explanation

Kathy's cakes are tasty. So, she should receive many cake orders. But the problem is they are expensive. So, that is the reason or the cause to get few orders.

Q: 12 Observe the following patterns of numbers.

- [2, 4, 5, 22]
- [6, 7, 8, 62]
- [8, 6, 5, A]

Based on the pattern, what is value of A ?

A. 53

✓ Correct Ans **B.** 38

C. 30

D. 48

Explanation

The fourth number within every bracket is made up with the first 3.

Last number = first number + second number x third number

So, A = 8 + 30 = 38.

Q: 13 John had just finished his Selective test. He told his friends :
"Some questions in the exam were difficult."

Which of the following has the **opposite meaning** of what John had said ?

A. Some questions in the exam were easy.

B. All the questions in the exam were difficult.

C. Some questions in the exam were not difficult.

✓ Correct Ans **D.** All the questions in the exam were easy.

Explanation

The opposite or the negation of "Some questions were difficult" is no question was difficult. Or we can say that, all the questions were easy.

Q: 14 **Statement** : "All engineering students should have a 6-month training experience before getting a job."

Neville and Smith are *providing their opinions* based on the above statement.

Neville: "Engineering students don't have practical skills before training."
Smith: "Training help students to gain some good knowledge and skills related to the industry."

Who is most correct according to statement ?

✓ Correct Ans **A.** Only Smith is correct

B. Only Neville is correct

C. Both are correct

D. Both are wrong

Explanation

The statement doesn't reveal anything about the previous practical skills of students. But every student should attend a 6-month training because before going to the industry. They should have at least some knowledge and experience about what they are going to do in the industry.

Q: 15 Students in group A are good football players and athletes. Students in group B are not into athletics but they are good at cricket. Group C students are good at all the sports except athletics. There are no other groups.

Which of the following are correct about the students?

A. Group A and C are only good at athletics

B. Only two groups are not good at cricket

C. Only one group has are good cricketers

✓ Correct Ans **D.** Only two groups have good football players

Explanation

Games that each group good at

- Group A → Football, athletics
- Group B → Cricket
- Group C → All except athletics

So, group A and C are good at football.

Q: 16 Karen is an electrical engineering student. During Lab time, he told is friend Rosy :
"If your circuit is not working, probably there is a short circuit, or the IC is blown."

But later Rosy found the error and Karen was proven wrong.
So, which is the most correct conclusion from the situation ?

A. There were no short circuits

B. The IC wasn't blown

✓ Correct Ans **C.** There were no short circuits, and the IC wasn't blown

D. There were no short circuits, but IC can be either blown or not.

Explanation

Karen is wrong. So, negation is correct. That means, circuit was not working properly, there were no short circuits, and the IC was not blown.

Q: 17 You are given a square shaped paper to make a hole in the middle as shown in the figure.

Minimally, how many times do you need to fold the paper to make the hole with a single cut?

A. 1

✓ Correct Ans **B.** 2

C. 3

D. 4

Explanation

Fold the sheet in half. Again, fold the sheet in half. Then cut near the middle point of the paper. You will get the above figure.

Q: 18 **Linda :** "Every photographer has an expensive camera. I know that Mary doesn't have an expensive camera."

Which one of the below may be Linda's conclusion ?

✓ Correct Ans **A.** Mary is not a photographer

B. If Mary buys an expensive camera, she will become a photographer

C. No conclusion can be made

D. None of the above

Explanation

If someone if a photographer, she has an expensive camera. From the contrapositive on the other hand, if someone doesn't have an expensive camera, he/she is not a photographer. So, Mary is not a photographer.

Q: 19 The following statements are true.

a. The square of X is 16
b. Y=0
c. L=2XY

When finding the value of L, which statement(s) is/are necessary?

A. a only

✓ Correct Ans **B.** b only

C. Neither a nor b

D. Both a and b

Explanation

Any number multiplied by 0 is 0; hence, there is no need to use statement a to find the value of L. B is the correct answer.

Q: 20 A dice has faces labeled 1 to 6. Which statement is necessarily true if 2 is adjacent to 1, 3, and 5?

A. 1 is opposite to 5

B. 2 is adjacent to 4

✓ Correct Ans **C.** 4 is adjacent to 6

D. 4 is opposite to 6

Explanation

If 2 is adjacent to 1, 3, and 5, then either 4 or 6 lies opposite to 2. Therefore, 4 and 6 can't be opposite each other. So, 4 and 6 can only be adjacent. Hence, C is correct.

Q: 21 James walked northwards for 20 meters. Upon turning left, he went for 40 meters. James then turned left again, walking for 20 meters again. Lastly, he turned right and walked for 20 meters. How far had he traveled from where his journey began?

A. 40

B. 100

C. 80

✓ Correct Ans **D.** 60

Explanation

The figure below shows mikes movement :

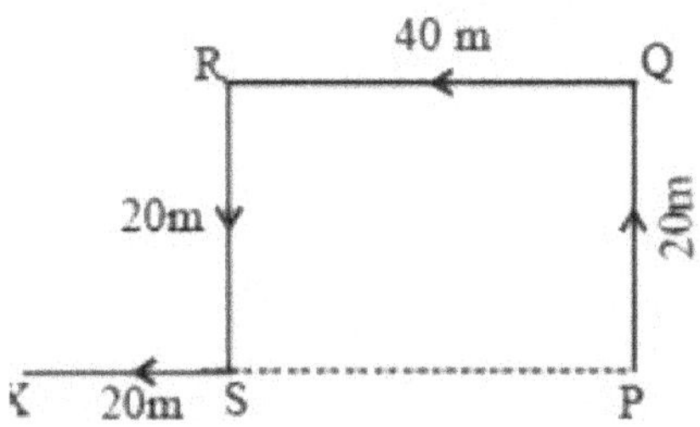

Since he started at P and stopped at X, we have 40 + 20 = 60 m.

Q: 22 Among A, B, C, D and E, A is taller than B but shorter than C. B is taller than only E . C is not the tallest.

Who among them will be in the middle if they stand in the order of their heights?

✓ Correct Ans **A.** A

B. C

C. B

D. E

Explanation

A is taller than B but shorter than C : C>A>B

B is taller than only E : C>A>B>E

C is not the tallest. D>C>A>B>E

Hence A is in the middle.

Q: 23 **Identify implicit assumption(s) behind the statement.**

Statement: "Those who are appearing for this examination for the first time, should be helped in filling up the forms". - An instruction to the invigilating staff.

Assumption I: The form is somewhat complicated.
Assumption II: Candidates can appear more than once for this examination.

A. Only Assumption I is implicit

B. Only Assumption II is implicit

C. Neither Assumption I nor II is implicit

✓ Correct Ans **D.** Both the assumptions are implicit.

Explanation

The statement mentions those who appearing first time, indicating second attempt is allowed. Also there is help available because filling the form is complicated.

Q: 24 ***If Sara enters her dream university, she will leave her country.***
If Sara leaves her country, she will have to cook for herself on her own.

If the above information is true, which of the following cannot be true?

A. Sara left her country, but she didn't enter her dream university

B. Sara cooked for herself on her own but she didn't enter her dream university

✓ Correct Ans **C.** Sara entered her dream university but she didn't cook for herself on her own

D. Sara cooked for herself on her own but she didn't leave her country.

Explanation

If Sara enters her dream university, she will leave her country, and then she have to cook for herself on her own. So option C is wrong

Q: 25 Mr. Leon promised Jason that he will deposit money to Jason's account before 4 p.m. but only if he gets some free time.

Jason: "Now it's 10 p.m. Mr. Leon still hasn't deposited money, he must not have gotten free time"

Nicole: "If he deposited money before 4 p.m. then definitely he has got the free time."

If the information box is true, whose reasoning is not correct?

A. Jason only

✓ Correct Ans **B.** Nicole only

C. Both Jason and Nicole

D. Neither Jason nor Nicole

Explanation

Mr. Leon will deposit money to Jason's account before 4 p.m. only if he gets some free time. So if he deposited the money, that means he must have got some free time. So Nicole is correct.

But there may be other reasons why he didn't deposit money. So Jason is wrong.

Q: 26 **Indian paradise flycatchers are beautiful bird whose male have elongated central tail feathers. Their heads are glossy black with a black crown and crest.**

Adam: "If a male bird has elongated central tail feathers, that bird must be an Indian paradise flycatcher"

Hanna: "If a bird's head is glossy black with a black crown and crest, it must be an Indian paradise flycatcher"

If the information is true, whose reasoning is correct?

A. Adam only

B. Hanna only

C. Both Adam and Hanna

✓ Correct Ans **D.** Neither Adam nor Hanna

Explanation

The fact that male Indian paradise flycatchers have elongated central tail feathers, doesn't mean any male bird who has elongated central tail feathers is an Indian paradise flycatcher.
Also, their heads are glossy black with a black crown and crest doesn't mean any bird whose heads are glossy black with a black crown and crest is an Indian paradise flycatcher.

Q: 27 ***Mr. Roger is an investigator. He knows about everyone, every criminal and notorious people in his area where he lives in. He is living there for 20 years. He has a special ability to memorize anyone's face or actions.***

Mr.Roger: "Just show me the CCTV video of the thief. I will tell exactly who is he by his movements!"

Which one of the following sentences, if true, shows the mistake Mr. Roger has made?

A. He can't memorize all the faces and movements of every person in the area for such a long time.

B. He can't remember every movement of a single person.

✓ Correct Ans **C.** He does not know about the thieves who do not live or came before into his area.

D. He can only give us predictions, he can't prove who is the thief.

Explanation

Mr. Roger knows about everyone, every criminal and notorious person in the area. But if the thief is not a person in the area he knows, then he can't tell exactly who is he by his movements

Q: 28 **Archaeopteryx is an iconic fossil, often thought of as the 'missing-link' between dinosaurs and birds. They are sometimes referred to by its German name, "Urvogel", an extinct genus of bird-like dinosaurs.**

A group of friends is visiting a huge forest with a guide.

Guide: "Don't go here and there alone as there might be very dangerous animals like bears and urvogels"

Which of the following shows what is wrong with the guide's advice?

A. If they are not allowed to walk freely, what is the use of coming to a forest, its better if they stay at home instead.

B. Walking in that dangerous forest cannot guarantee safety.

✓ Correct Ans **C.** Archaeopteryx is an extinct animal so they can't be in that forest.

D. There are many more dangerous things in the forest.

Explanation

As in the given info box Archaeopteryx, also known as Urvogel is an extinct animal from the dinosaur era. So they can't be there in that forest.

Q: 29 **Nitrogen is an essential element for plant function and growth. Plants get the nitrogen that they need from the soil. Nepenthes is an insectivorous plant. Insectivorous plants have special leaves which can trap insects and kill them to derive nitrogen from the insects.**

If the information is correct, which of the following statement is correct?

✓ Correct Ans **A.** Insectivorous plant might grow in nitrogen-deficient soil.

B. Nitrogen is important for every plant, therefore Nepenthes can't grow in nitrogen-deficient soil.

C. If a plant has special leaves which can trap insects and kill them that must be a Nepenthes plant.

D. As nepenthes is an insectivorous plant it does not need nitrogen to grow.

Explanation

As nepenthes use their special leaves to trap insects and kill them and derive nitrogen from them, nepenthes can grow up in nitrogen-deficient soil. This is answer A.

Q: 30 Global warming and air pollution are serious threats to life on earth. Vehicle engines are the main reason for both air pollution and increasing global warming. If we reduce using vehicles by a significant amount, we can reduce air pollution too. For that, the government has to legislate laws and rules to completely stop private vehicle production, like cars, vans, jeeps, etc., and encourage people to use public transportation. Along with that, the government has to introduce a system to recycle private vehicles and manufacture bicycles, which are friendly for both environment and health. So people can travel short distances by bicycle and long distances by bus or train or any other public transport method.

Which of the following sentences most weakens the above argument?

✓ Correct Ans **A.** The air is polluted by public transport methods too.

B. Recycling private vehicles and manufacturing bicycles is a waste of money.

C. Private vehicles can be colourful.

D. Cycling stimulates and improves your heart, lungs and circulation, reducing your risk of cardiovascular diseases.

Explanation

The argument can be weakened by telling even public transport contributes to pollution.

Q: 31 The time remaining to finish 100 pages of the book is one week (7 days). Which of the following will ***surely*** make you miss your deadline?

A. Reading 20 pages per day for five days

B. Reading 25 pages every other day

C. Reading 30 pages per day for the last three days

✓ Correct Ans **D.** Reading 10 pages each day for the whole week

Explanation

Reading 10 pages each day for 7 days would not allow to complete 100 pages, for sure.

Q: 32 Incandescent bulbs are having a tungsten coil. Also, their power consumption is high.
However, LED bulbs are totally the opposite.

Based on above information, LED bulbs ....

A. Do not have a tungsten coil.

B. Do not have a tungsten coil but the power consumption is high.

C. The power consumption is low but they don't have a tungsten coil.

✓ Correct Ans **D.** Don't have a tungsten coil or the power consumption is low.

Explanation

If a bulb is incandescent, it has a tungsten coil and it consumes more power. Given that, LED bulbs are opposite then that means: LED bulbs don't have a tungsten coil or they have a lower power consumption.

Q: 33 **Carpenter**: "If it doesn't rain I can finish your work by next Monday."

Robin: "Are you sure? It is very hard to believe."

What does Robin think?

A. If it doesn't rain the Carpenter will finish the work by next Monday.

B. It will rain and Carpenter will finish the work by next Monday.

C. Carpenter will finish the work by this Monday.

✓ Correct Ans **D.** It won't rain and Carpenter will not finish the work by next Monday.

Explanation

Robin thinks what the carpenter said was not true. That means, she thinks the negation of what Carpenter said was true. The negation is, it won't rain and he cannot finish the work by next Monday.

Q: 34 If there is heavy rainfall for two days, there is a risk of flood in Kevin's hometown.
If there is a risk of flood, people have to move to a safe place.

Kevin said it was heavily raining there since the day before yesterday. Which of the following is correct based on information above?

✓ Correct Ans **A.** Kevin's family has to move to a safer place.

B. Kevin's family doesn't need to move to another place.

C. No conclusion can be made.

D. None of the above.

Explanation

By combining the statements we can find - "if there is heavy rainfall for two days, then people have to move to a safe place". Given that it is heavily raining in Kevin's hometown, they have to move to a safe place.

Q: 35 I said : "*If it's raining, we cannot play cricket. If the bat is broken, we cannot play cricket either.*"

Rickey said he bought a new bat today.

Which of the following is correct about the given scenario?

A. We cannot play cricket.

✓ Correct Ans **B.** Rain is the only reason we won't play cricket.

C. We can play cricket even if it rains.

D. None of the above.

Explanation

From the given statements we can tell that, if it rains or the bat is broken, then we cannot play cricket. Since Ricky bought a new bat, we know that the bat is not broken. So, now there is one reason to decide whether we can play cricket or not. That is the rain.

Q: 36 **Statement** : If it is spring, farmers start farming.

Based on the **above statement alone,** which ones of the below must be true ?

A. If it is not the spring, farmers don't farm.

B. If farmers are farming, it is spring.

C. If farmers are not farming, it is the winter.

✓ Correct Ans **D.** If farmers are not farming, it may not be the spring.

Explanation

The contrapositive of the statement has the same meaning. That is, "If the farmers don't start farming, then it is not spring."

Q: 37 *All Italian restaurants have tasty food.*
James had a meal in ABC restaurant and he said it was tasty.

What can be concluded about ABC resturant based on above two statements ?

A. ABC is an Italian restaurant.

B. ABC isn't an Italian restaurant.

✓ Correct Ans **C.** Cannot decide.

D. ABC resturant is closed.

Explanation

Given that, if a restaurant is an Italian one, they have tasty food. However, that doesn't always work in the other direction. So, although we know the meal is tasty we cannot conclude that it was an Italian restaurant.

Q: 38 *If the sun is so shiny and the rainfall is very low, the lakes will dry up.*
The sun is so shiny these days.

Which of the following is the most correct conclusion regarding the above scenario?

A. If the rainfall is very high the lakes won't dry up.

B. The lakes will dry up.

C. The lakes won't dry up.

✓ Correct Ans **D.** If the rainfall is very low the lakes will dry up.

Explanation

Given that, if the sun is so shiny and the rainfall is very low, then the lakes will dry up. It is also given that the sun is so shiny these days. So, if the rainfall is also low, then the lakes will dry up.

Q: 39 If the meaning of "(Thomas)@ (restaurant) * (Kamala) = (dinner) $ (7 O' clock)" is "Thomas went to a restaurant with Kamala for dinner before 7 O' clock", what is the meaning of "(Michael) @ (cinema) $ (film starts)"?

A. Michael goes to the cinema before it starts.

✓ Correct Ans **B.** Michael went to the cinema before the film starts.

C. Michael went to the cinema and watched a movie.

D. Michael went to the cinema with the film starts.

Explanation

(Michael) @ (cinema) -> Michael went to the cinema

$ (film starts) -> before the film starts

Combine them to get the answer.

Q: 40 **Sam :** *If a country is economically stable, people have money.*

Ram : *If people have money, they celebrate many festivals.*

Sam and Ram are both correct. However, in some countries, people doesn't celebrate many festivals. That means...

A. Some countries are economically stable.

✓ Correct Ans **B.** Some countries are economically unstable.

C. Cannot decide.

D. Economy is not an important aspect for a country.

Explanation

Applying hypothetical syllogism, we get “If a country is economically stable, people celebrate festivals”. On the contrapositive, “If people don't celebrate festivals, then the country is economically unstable”. In some countries, people doesn't celebrate festivals. So, that means some countries are economically unstable.

Answers And Solutions
Learnikx Education

Q: 1 According to Extract A, what is one characteristic that sets cats apart from dogs?

A. Their ability to communicate through vocalizations.

✓ Correct Ans **B.** Their independent nature and preference for personal space.

C. Their instinctive hunting skills.

D. Their reliance on scent marking for communication.

Explanation

According to the 4th paragraph, cats are known for their independent nature and preference for personal space. Unlike dogs, they value their autonomy and tend to exhibit more self-reliant behavior. While they form deep bonds with their human companions, they often require and appreciate their personal space.

Q: 2 According to Extract A, what is the importance of providing an enriched environment for cats?

A. It helps maintain their physical health.

B. It fosters a stronger bond with their human companions.

✓ Correct Ans **C.** It allows them to engage in natural behaviors and mental stimulation.

D. It ensures they have a clean litter box and fresh water.

Explanation

According to the 5th paragraph, providing an enriched environment for cats is important as it allows them to engage in natural behaviors and mental stimulation. Having access to scratching posts, interactive toys, and vertical spaces helps them exercise, maintain healthy claws, and prevent boredom. An enriched environment promotes their overall well-being and happiness.

Q: 3 According to Extract A, what is a responsible practice of cat ownership?

A. Regular playtime and interactive sessions.

B. Feeding them a balanced diet and providing veterinary care.

C. Ensuring access to clean water and a clean litter box.

✓ Correct Ans **D.** All of the above.

Explanation

According to Extract A, responsible cat ownership involves a combination of practices, including regular playtime and interactive sessions, providing a balanced diet and veterinary care, as well as ensuring access to clean water and a clean litter box. All these aspects contribute to the overall health and happiness of cats.

Q: 4 According to Extract B, how did the cat Billy help save his owner's life?

A. He called out for help when the owner suffered a heart attack.

✓ Correct Ans **B.** He woke up the owner by pounding his paws on her chest.

C. He ran to get the owner's mother for assistance.

D. He alerted the owner by jumping from the turret.

Explanation

According to Extract B, the cat Billy saved his owner's life by waking her up when she suffered a heart attack in her sleep. He accomplished this by pounding his paws on her chest, which alerted the owner to her condition and allowed her to seek help.

Q: 5 According to Extract B, what did the 2020 study by Arizona State University researchers reveal about dog's reactions in a distressing situation?

A. Dogs were trained to rescue their owners in a locked box.

✓ Correct Ans **B.** Dogs showed a physiological response to their distressed owner's cries.

C. Dogs were indifferent to their owner's calls for help.

D. Dogs lacked the intelligence to figure out how to open a locked box.

Explanation

According to Extract B, the study showed that dogs demonstrated a desire to help their owners and were intelligent enough to figure out the actions required to rescue them from a seemingly trapped situation. The dogs also exhibited a physiological response to their distressed owners' cries, indicating their concern and emotional connection.

Q: 6 According to Extract B, what did the study from Oregon State University conclude about the attachment styles of cats?

A. Cats are less attached to their owners compared to dogs.

B. Cats only exhibit attachment behaviors when they require food or attention.

✓ Correct Ans **C.** Cats show similar attachment styles to dogs and human infants.

D. Cats are more likely to display insecure attachments than dogs.

Explanation

According to Extract B, the study concluded that cats display attachment styles similar to dogs and human infants. The experiment's results showed that a significant percentage of the cats exhibited secure attachments to their owners, demonstrating trust, seeking protection, and expecting their owners to return. This finding challenges the notion that cats are less attached to their owners compared to dogs and suggests that cats care about their human caregivers just as much.

Q: 7 What are medium-chain triglycerides (MCTs) found in coconuts known for?

A. Supporting brain function

B. Aiding in digestion

C. Boosting metabolism

✓ Correct Ans **D.** All of the above

Explanation

The article mentions that MCTs found in coconuts have been linked to improved brain function, enhanced digestion, and increased metabolism.

Q: 8 How is coconut oil commonly used in skincare?

A. Hair conditioner

B. Moisturizer

C. Makeup remover

✓ Correct Ans **D.** All of the above

Explanation

According to the article, coconut oil can be used as a hair conditioner, moisturizer, and makeup remover in skincare routines.

Q: 9 What makes coconuts an environmentally friendly crop?

A. They require minimal maintenance.

B. Different parts of the tree can be utilized for various purposes.

C. They contribute to reduced environmental impact.

✓ Correct Ans **D.** All of the above

Explanation

The article explains that coconuts are environmentally friendly because they require minimal maintenance, different parts of the tree can be utilized, and they contribute to reduced environmental impact.

Q: 10 In the second stanza, what do the "soft, white hands" of the waves symbolize?

A. The welcoming embrace of the sea
B. The traveler's footprints in the sand
C. The darkness of the evening
✓ Correct Ans D. The passage of time

Explanation

The "soft, white hands" of the waves symbolize the erasure of the traveler's footprints in the sand, representing the passage of time and impermanence.

Q: 11 What is the deeper meaning or theme of the poem as reflected in the repeated line "And the tide rises, the tide falls"?

✓ Correct Ans A. The cyclical nature of life and death
B. The unpredictability of nature
C. The eternal beauty of the sea
D. The traveler's persistence in his journey

Explanation

The repeated line "And the tide rises, the tide falls" symbolizes the cyclical nature of life and death, emphasizing the transient nature of human existence.

Q: 12 How does the poet create a sense of melancholy in the poem?

A. Through descriptions of joyful moments
B. Through the use of vivid colors
C. Through the repetition of the tide's movement
✓ Correct Ans D. Through the traveler's inability to return

Explanation

The poet creates a sense of melancholy through the traveler's inability to return, suggesting a sense of finality and loss.

Q: 13 What effect does the repetition of the line "And the tide rises, the tide falls" have **on the poem's structure**?

✓ Correct Ans A. It adds a sense of rhythm and continuity.
B. It creates a disjointed and chaotic structure.
C. It emphasizes the beauty of sea.
D. It signifies the changing of seasons.

Explanation

The repetition of the line "And the tide rises, the tide falls" adds a sense of rhythm and continuity, reinforcing the cyclical theme.

Q: 14 Which stanza of the poem portrays a sense of renewal or rebirth?

A. The first stanza
B. The second stanza
✓ Correct Ans C. The third stanza
D. None of the stanzas

Explanation

The third stanza portrays a sense of renewal or rebirth as it describes the morning breaking and the day returning.

Q: 15 Which of the extracts refers to the science and art of cultivating plants?

A. Extract A
B. Extract B
C. Extract C
✓ Correct Ans D. Extract D

Explanation

The paragraph mentions Horticulture, the science and art of cultivating plants.

Q: 16 Which of the extracts refers to a field that encompasses the exploration of animal behavior?

A. Extract A
✓ Correct Ans B. Extract B
C. Extract C
D. Extract D

Explanation

The paragraph states that Zoology, the scientific study of animals, is a diverse and captivating field that encompasses the exploration of animal behavior, physiology, and evolution.

Q: 17 Which of the extracts refers to the understanding of the role of plants in sustaining ecosystems?

✓ Correct Ans A. Extract A
B. Extract B
C. Extract C
D. Extract D

Explanation

Extract A mentions that Botany is essential for understanding the role of plants in sustaining ecosystems.

Q: 18 Which of the extracts talks about the conservation of endangered species?

A. Extract A

✓ Correct Ans **B.** Extract B

C. Extract C

D. Extract D

Explanation

Extract B mentions that Zoology plays a significant role in conservation efforts, as scientists work to protect endangered species and their habitats.

Q: 19 Which one fits in (1) ?

A. Sentence A

B. Sentence B

✓ Correct Ans **C.** Sentence G

D. Sentence H

Explanation

The essay in on microorganisms, therefore, the correct sentence should be a introduction to it.

Q: 20 Which one fits in (2) ?

✓ Correct Ans **A.** Sentence A

B. Sentence B

C. Sentence C

D. Sentence D

Explanation

The next line states about the size of the creatures. So, the correct answer should have the word "size" for the easiness in selecting the correct answer.

Q: 21 Which one fits in (3) ?

A. Sentence A

B. Sentence D

✓ Correct Ans **C.** Sentence B

D. Sentence E

Explanation

The start of the paragraph says that microbes are essential for maintaining the eco system. Therefore, the correct answer shook be related to the environment.

Q: 22 Which one fits in (4) ?

A. Sentence A

B. Sentence B

C. Sentence H

✓ Correct Ans **D.** Sentence E

Explanation

The next parts of the paragraph explain about the industrial used of the microbes.

Q: 23 Which one fits in (5) ?

A. Sentence A

B. Sentence B

✓ Correct Ans **C.** Sentence C

D. Sentence D

Explanation

The paragraph explains about the harmful cases for humans. So that the correct s entice should be related to the harmful ways of microbes to the humans.

Q: 24 Which one fits in (6) ?

A. Sentence A

B. Sentence B

C. Sentence C

✓ Correct Ans **D.** Sentence D

Explanation

The next parts of the paragraph explain about the scientific studies related to the microbes. Therefore, the correct answer should be an introduction to such researches.

Q: 25 Which one fits in (7) ?

A. Sentence A

B. Sentence G

✓ Correct Ans **C.** Sentence H

D. Sentence C

Explanation

The next parts of the paragraph have the word "microbiome". Thus, the correct sentence should have that word.

Q: 26 What is the tone of the poem's opening lines?

A. Joyful

✓ Correct Ans B. Despairing

C. Indifferent

D. Ironic

Explanation

The tone in the opening lines is despairing as the speaker reflects on his loss of sight and its impact.

Q: 27 In the poem, what does the speaker consider a "Talent"?

✓ Correct Ans A. His ability to write poetry

B. His physical strength

C. His knowledge of the world

D. His wealth and possessions

Explanation

The speaker considers his ability to write poetry a "Talent."

Q: 28 What is the deeper meaning or message conveyed by the poem?

A. The importance of worldly success

B. The inevitability of physical decline

✓ Correct Ans C. The value of patience and acceptance in serving God

D. The futility of serving God without expecting rewards

Explanation

The poem's deeper message is about the value of patience and acceptance in serving God despite personal limitations.

Q: 29 Which phrase from the poem suggests that God does not require human efforts or offerings?

A. "my light is spent"

✓ Correct Ans B. "God doth not need"

C. "they serve him best"

D. "his mild yoke"

Explanation

The phrase "God doth not need" suggests that God does not require human efforts or offerings.

Q: 30 What does the speaker mean by "his mild yoke" in the poem?

✓ Correct Ans **A.** God's gentle guidance

B. God's heavy burden

C. God's strict rules

D. Our material wants

Explanation

"His mild yoke" refers to God's gentle and forgiving guidance.

Q: 1 I am a prime number between 10 and 40. I am a factor of 111.

What number am I ?

A. 35

B. 36

✓ Correct Ans **C.** 37

D. 38

Explanation

Factors of 111 = 1,3,37,111.

The number is 37 which is between 10 and 40.

Q: 2 Noah bought 231 eggs. Out of this, he sold $^{9}/_{11}$ of eggs and $^{2}/_{7}$ of the **remaining eggs** were broken. How many eggs will Noah have at the end?

✓ Correct Ans **A.** 30

B. 42

C. 36

D. 41

Explanation

Sold = (231 x 9) / 11 = 189
Remaining = 42
Broken = 42 x $^{2}/_{7}$ = 12

Left = 42 - 12 = 30

Q: 3 At Fresh Bread and Cakes bakery, the price of one fresh loaf of bread is $1. However, if there are leftovers, each of these loaves is sold $0.40 cheaper on the next day. The table below shows the number of loaves baked and sold over two days. On August 2nd, all loaves were sold. Find the total earnings for the two days.

Days	Number of loaves baked	Total number sold
August 1st	90	62
August 2nd	90	118

✓ Correct Ans **A.** $168.80

B. $170.00

C. $124.30

D. $190.00

Explanation

August 1st
62 fresh loaves = 62 x $1 = $62

August 2nd
Left over = 90 - 62 = 28
Money from left = 28 x ($1.00 - 0.40) = $16.80
Fresh loaves's money = 90 x $1 = $90
Total earned - $(62 + 90 + 16.80) = $168.80
The total earning is $168.80.

Q: 4 John and Peter shared some marbles in the ratio of 7:5. John had 24 more marbles than Peter. How many marbles did Peter have?

A. 12

✓ Correct Ans **B.** 60

C. 84

D. 144

Explanation

John has 7n, Peter has 5n.
7n - 5n = 2n = 24; n=12; 5n = 60

Q: 5 Sam spent $^1/_4$ of his money at recess. He spent $^1/_6$ of his **remaining money** on lunch. What fraction of his money was left in the end?

A. $^1/_8$

✓ Correct Ans B. $^5/_8$

C. $^5/_{12}$

D. $^7/_{12}$

Explanation

Spent on recess : $^1/_4$
Remaining = $1 - {}^1/_4 = {}^3/_4$

Lunch = 3/4 x 1/6 = 3/24 = 1/8

Left = $1 - {}^1/_4 - {}^1/_8 = {}^{(8-2-1)}/_8 = {}^5/_8$

Q: 6 A rope of length 72 cm was cut into three pieces. The first piece was three times as long as the second piece. The second piece was twice as long as the third piece.

How long was the second piece?

A. 12 cm

✓ Correct Ans B. 16 cm

C. 18 cm

D. 24 cm

Explanation

Third piece = n
Second piece = 2n
First piece = 2n x 3 = 6n
Total = 6n + 2n + n = 9n
9n = 72; n = 8

So, second piece = 2n = 16 cm

Q: 7 In a garden there is a row of trees, one tree is fifth from either ends of this row.

How many trees are there in the row?

A. 5

B. 8

C. 10

✓ Correct Ans D. 9

Explanation

The tree in middle is counted twice.
Total number of trees = 5 + 5 - 1 = 9

Q: 8

Linda was given 100 minutes to complete a test. She took $\frac{3}{4}$ of the time to complete the test and she stopped writing after that. She started doing the test at 14:50.

At what time did she stop writing?

A. 4:00 PM

✓ Correct Ans **B.** 4:05 PM

C. 5:00 PM

D. 6:05 PM

Explanation

4 units = 100 min
1 unit = 100 min ÷ 4 = 25 min
3 units = 25 min x 3 = 75 min = 1h 15min
2:50 pm + 1h 15min = 4.05 pm

Q: 9

A photo frame measures 15 centimeters by 10 centimeters. A photograph is placed in it, leaving a border of 2 centimeters wide all around it.

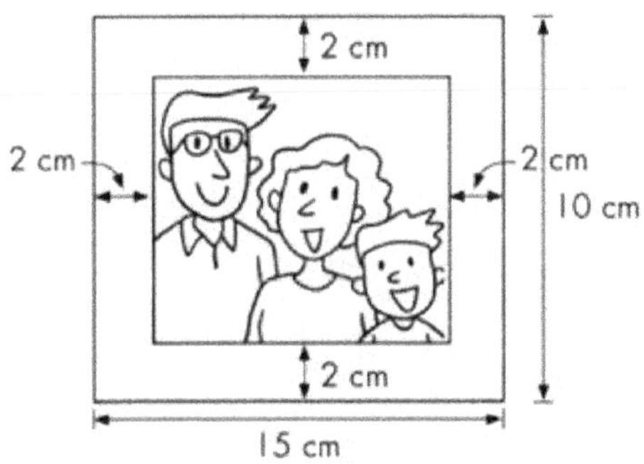

What is the area of the border?

A. 80 sq cm

✓ Correct Ans **B.** 84 sq cm

C. 100 sq cm

D. 160 sq cm

Explanation

Area of photo frame = 15 x 10 = I50
Length of photograph = 15 - 2 - 2 = 11
Breadth of photograph = 10 - 2 - 2 = 6
Area of photograph → 11 x 6 = 66
Area of border = 150 - 66 = 84
The area of the border is 84 square centimeters.

Q: 10 Grandpa had fewer than 20 finger puppets. If he gave each of his grandchildren 3 finger puppets, he would have 2 finger puppets left. If he gave each grandchild 4 finger puppets, he would be short of 3 finger puppets.

How many finger puppets did he have?

A. 10

B. 11

✓ Correct Ans **C.** 17

D. 20

Explanation

Multiples of 3: 3, 6, 9, 12, 15, 18
+2 : 5 , 8, 11 , 14 ,**17** , 20

Multiples of 4: 4, 8, 12, 16, 20, 24
-3 : 1 , 5, , 9, 13, **17**, 21

Grandpa had 5 or 17 finger puppets. If he had 5 finger puppets, he could only give 1 grandchild 3 finger puppets with 2 finger puppets left. He was able to give 3 finger puppets to each of his grandchildren so he should have more than 5 finger puppets. So, Grandpa had 17 finger puppets.

Q: 11 5 notebooks, 2 pencils and 2 markers cost $10. 1 pencil and 1 marker cost $2.
How much does 1 notebook cost?

A. $6.20

✓ Correct Ans **B.** $1.20

C. $4.00

D. $10.00

Explanation

1p + 1m = $2
2p + 2m = $4
5n + 2p + 2m = $10

Subtracting :
5n = $10 - $4 = $6
1n = $6 ÷ 5 = $1.20

Q: 12 At a fruit shop, mangoes are sold at 5 for $9 while apples are sold at 7 for $8. Mr Rao bought an equal number of mangoes and apples. He spent $92 more on the mangoes.
How many apples did Mr Rao buy?

A. 150

B. 5

✓ Correct Ans **C.** 140

D. 144

Explanation

Mangoes	Apples
5 for $9	7 for $8
35 for $63	35 for $40

$63 - $40=$23 (1 group difference)

How many groups ?
92 ÷ 23=4 (groups)

4 groups have 4 X 35=140 apples.

Q: 13 The table below shows the opening hours of a clinic in a day.

Opening Hours
8.30 a.m. - 12.45 p.m.
2.00 p.m. - 4.30 p.m.
7.00 p.m. - 8.30 p.m.

How long is the clinic open per day ? Leave your answer in hours and minutes.

A. 10h 30min

B. 2h 10min

C. 6h 14min

✓ Correct Ans **D.** 8h 15min

Explanation

From 8.30 a.m. to 12.45 p.m. → 4h 15min = 255min
From 2.00 p.m. to 4.30 p.m. → 2h 30min = 150min
From 7.00p.m. to 8.30p.m.→1h 30min = 90min
255 + 150 + 90 = 495min
495min = 8h 15min

Q: 14

A rectangular tank measuring 150 cm by 70 cm by 85 cm high was $\frac{2}{5}$ filled with water. Mrs Sam poured 12 pails of water, each containing the same volume of water, into the tank. In the end, there was 615 L of water in the tank. What was the volume of water in each pail?
Give your answer in litres.

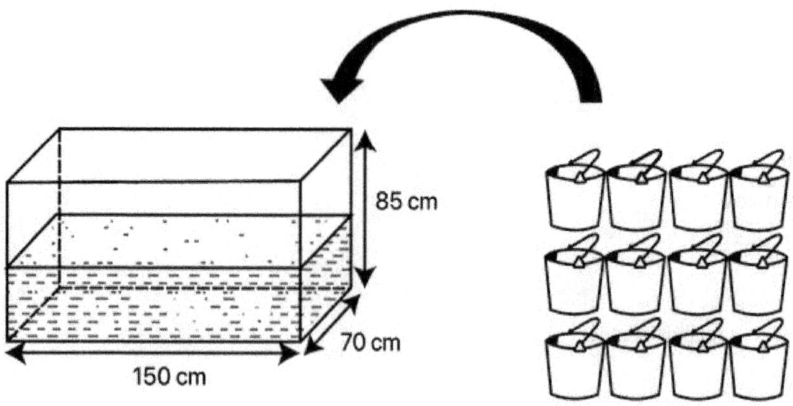

✓ Correct Ans **A.** 21.5L

B. 23.6L

C. 33.8L

D. 22.9L

Explanation

1000 cu cm = 1 L
Height is = $^2/_5$ of 85 = 34
Volume = 150 x 70 x 34 = 357000 cu cm = 357 L

How much poured = 615 - 357 = 258 L

Each pail = 258/12 = 21.5 L

Q: 15 Ravi has some marbles. If he gives 7 marbles to each of his friends, he will have 4 marbles left. If he gives each of his friends 9 marbles, he will need another 2 more marbles.
How many marbles does he have?

A. 39

✓ Correct Ans **B.** 25

C. 26

D. 90

Explanation

Let's make an organized list.
Multiple of 7 → 7 14 21 28 35
+4 → 11 18 {25} 32 39

Multiple of 9 → 9 18 27 36 45
-2 → 7 16 {25} 34 43

25 is the common number.

Q: 16 The mass of a glass bottle filled with 10 identical erasers is 1200 g. The mass of the same bottle when filled with 6 such erasers is 1040 g. What is the mass of each eraser?

A. 44g

B. 49g

C. 45g

✓ Correct Ans **D.** 40g

Explanation

Mass of 4 erasers → 1200 - 1040 = 160g
Mass of 1 eraser → 160 ÷ 4 = 40g

Q: 17

Mr Tan made some fish balls. He sold $\frac{3}{4}$ of the fish balls in the morning and $\frac{1}{3}$ of the remaining fish balls in the afternoon. He made another 506 fish balls and had twice the number of fish balls he made at first.
How many fish balls did he make at first?

✓ Correct Ans **A.** 276

B. 65

C. 287

D. 90

Explanation

Assume at first there were 12 parts.
Morning sold = 9 parts
Left = 3 parts, sold 1 part more
Total unsold = 2 parts

2 parts + 506 = 24 parts
1 part = 23
12 parts = 23 x 12 = 276

Q: 18

A total of $2,500 is shared between 3 pupils, Zay, Jack, and Rose. Rose gets $625. Zay gets 20% more money than Rose. Calculate the money that Jack receives.

✓ Correct Ans **A.** $ 1,125

B. $ 1,115

C. $ 1,525

D. $ 1,365

Explanation

Rose = 625
Zay : 625 x 20% = 125 ; 625 + 125 = 750
Jack = 2,500 – 625 – 750 = 1,125

Q: 19 The table shows the symbols and their values.

Symbol	Value
	35
	65
	37
	102
	65

Which of the followings is correct?

A. 

B. 

✓ Correct Ans C. 

D. 

Explanation

A. 65 – 102 = 37, correct answer = -37
B. 37 – 65 = 35 , correct answer = -28
C. 102 – 37 = 65, correct answer
D. 65 – 65 = 102, correct answer = 0

Q: 20 The bar chart below shows the number of people who likes a variety of fruits.

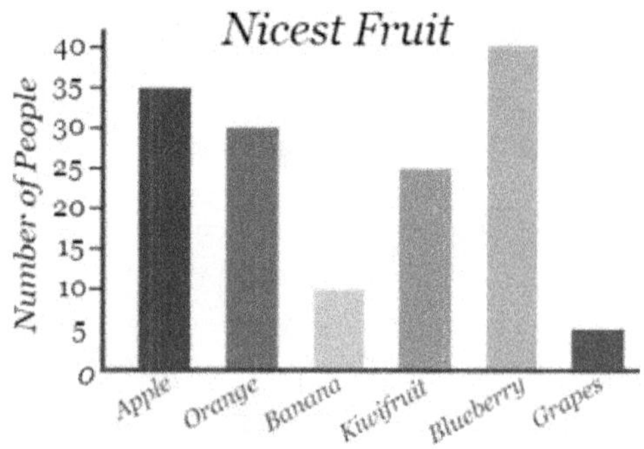

Which of the following is a correct statement?

A. The number of people who like bananas and apples is 50.

B. The number of people who like grapes and oranges is 45.

C. The number of people who like kiwifruit and blueberry is 75.

✓ Correct Ans **D.** The number of people who like orange and blueberry is 70.

Explanation

A. The number of people who like bananas and apples is 50 = wrong, correct is 45
B. The number of people who like grapes and oranges is 45 = wrong, correct is 35
C. The number of people who like kiwifruit and blueberry is 75 = wrong, correct is 65
D. The number of people who like orange and blueberry is 70 = correct statement

Q: 21 The table shows 4 colors of pens in a box.

Color	Pens
Purple	35%
Red	25%
Blue	
Black	10%

There were 7,000 pens in a box. Half of the blue is sold. How many **blue pens are remaining unsold** ?

A. 1,527

B. 1,505

C. 1,023

✓ Correct Ans **D.** 1,050

Explanation

7,000 x 35% = 2,450
7,000 x 25% = 1,750
7,000 x 10% = 700
7,000 – (700 + 1750 + 2450) = 2,100 ÷ 2 = 1,050

Q: 22 The pictures show a surface area of a shape.

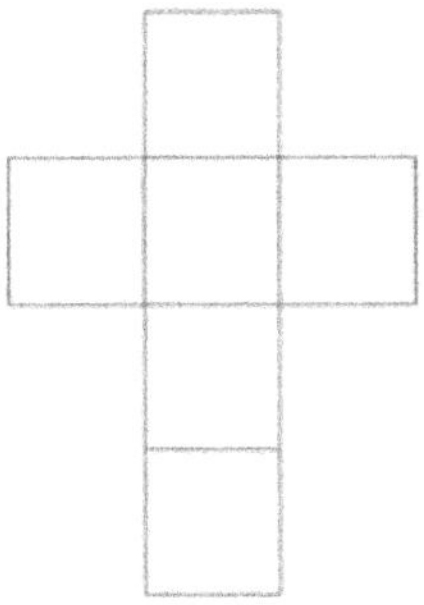

What shape will appear when the surface area got folded?

✓ Correct Ans **A.** Cube

B. Pyramid

C. Cone

D. Cylinder

Explanation

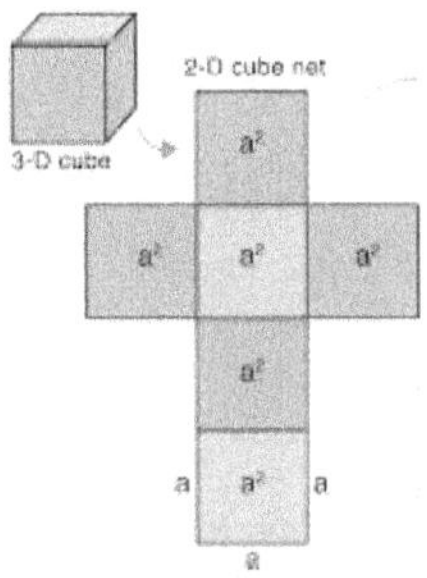

Q: 23 In a classroom activity, 16 pupils stood in a straight row with equal distances between each other. The distance between the 5th pupil and the 10th pupil was 11.95 m.
What was the distance between the first pupil and the last pupil?

A. 38.35 m

B. 36 m

✓ Correct Ans **C.** 35.85 m

D. 37.5 m

Explanation

For 5 gaps = 11.95 m
For 15 gaps = 11.95 x 3 = 35.85 m

Q: 24 The table below shows the charges of an automated bicycle rental system in city.

Rental charges	
For the first hour	\$1.50
For every subsequent $\frac{1}{2}$ hour or part thereof	\$0.50

How much must Michelle pay to rent the bicycle from 1.30 p.m. to 4.15 p.m.?

✓ Correct Ans **A.** \$3.50

B. \$3.00

C. \$4.50

D. \$5.00

Explanation

\$1.50 + (4 x \$0.50) = \$3.50

Q: 25 Ali, Ben and Carson shared some sweets. The ratio of the number of sweets Ali had to the total number of sweets Ben and Carson had is 3 : 8. Ali had the same number of sweets as Ben and 36 fewer sweets than Carson.
What is the ratio of the number of sweets Ali had to the number of sweets Ben had to the number of sweets Carson had?

A. 3:3:8

B. 3:5:3

C. 3:3:3

✓ Correct Ans **D.** 3:3:5

Explanation

There are extra information in the problem.
Ali = 3n
Ben + Carson = 8n
Ben = 3n; Ali = 8n - 3n = 5n
Ali : Ben : Carson = 3n:3n:5n = 3:3:5

Q: 26 The average test score of a group of students was 72 marks. After three students who scored an average of 85 marks left the group, the average score of the remaining students became 69 marks. How many students were there in the group at first?

A. 76

B. 156

C. 36

✓ Correct Ans **D.** 16

Explanation

There are total 'n' students.
Total score = 72xn
Left = 85x3 = 255
Remaining score = 72n - 255
72n - 255 = (n-3)x69 = 69n - 207
3n = 255 - 207 = 48; n = 16

Q: 27 Which of the following statements correctly describes the number of edges, vertices, and faces of a cube?

A. 6 edges, 8 vertices, 6 faces

B. 8 edges, 6 vertices, 12 faces

✓ Correct Ans **C.** 12 edges, 8 vertices, 6 faces

D. 4 edges, 6 vertices, 4 faces

Explanation

A cube is a three-dimensional solid shape with six square faces.

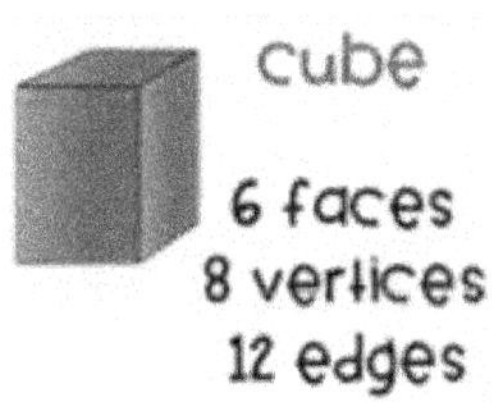

Q: 28 Table below shows the number of phones owned by students in school S.

Phone	Number of phones
Samsung	850
Huawei	60% of Samsung
Apple	561 more than Huawei

Calculate the total number of phones owned by the students.

A. 3,410

B. 8,240

✓ Correct Ans **C.** 2,431

D. 2,410

Explanation

Huawei = 850 x 60% = 510 , Apple = 510 + 561 = 1,071. So, total = 850 + 510 + 1,071 = 2,431

Q: 29 When Ahmed walked through a shopping mall, he recorded the number of customers in each shop. The data is displayed in a dot plot.

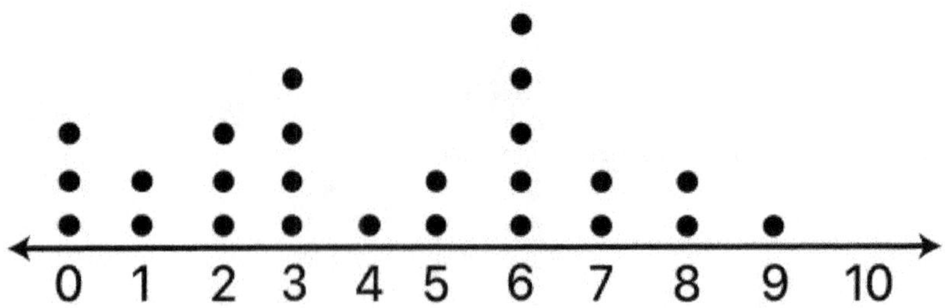

What percentage of shops had at least 5 customers?

A. 25%

✓ Correct Ans **B.** 48%

C. 33%

D. 44%

Explanation

Percentage = (Number of shops with at least 5 customers / Total number of shops) x 100 = $^{12}/_{25}$ x 100 = 48%

Q: 30 In a mixture of 60 litres, the ratio of milk and water is 2:1. If the ratio is to be 1:2, then the quantity of water to be further added is ..

A. 20 L

B. 30 L

C. 50 L

✓ Correct Ans **D.** 60 L

Explanation

Milk = 60 x $^{2}/_{3}$ = 40
Water = 60 – 40 = 20
Water to be added is 'x'.
$^{40}/_{(20 + x)} = ^{1}/_{2}$
20 + x = 80 ; x = 60

Q: 31 Sofia and her friends ordered 3 large pan pizza. Each pizza was cut into 8 pieces.

Iris ate 5 pieces.
Luna ate $^{1}/_{6}{}^{th}$ of the total.
Eloise ate $^{1}/_{3}{}^{rd}$ of the total, and Sofia ate the remaining.
Who ate most?

A. Iris

B. Luna

✓ Correct Ans **C.** Eloise

D. Sofia

Explanation

Total no. of pieces = 8 ×3 = 24
No. of pieces Luna ate = 24 ×$^{1}/_{6}$ = 4
No. of pieces Eloise ate = 24 ×$^{1}/_{3}$ = 8
No. of pieces Sofia ate = 24 – (8+4+5) = 24-17 = 7
Eloise has eaten the most.

Q: 32 This chart denotes favourite ice-cream falvours of people.

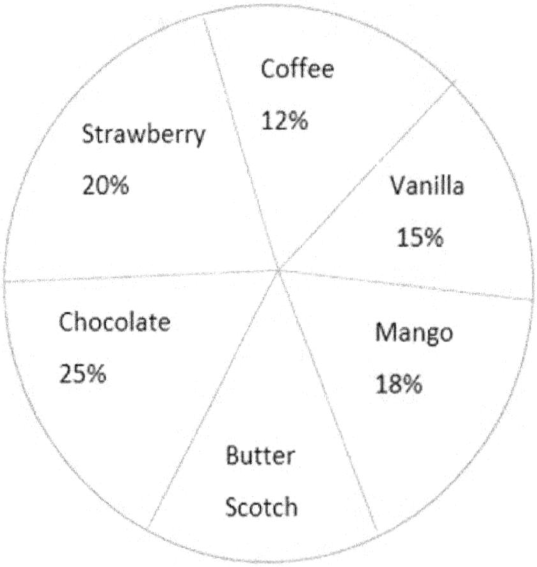

If 735 people like vanilla flavour, how many people like mango flavour?

A. 752

B. 798

C. 844

✓ Correct Ans **D.** 882

Explanation

735 = 15%

1% = $^{735}/_{15}$ = 49

Percentage for mango flavour = 18%

No. of people who like mango flavour = 49 ×18 = 882

Q: 33 $\angle X = 52°$. $\angle Y$ is twice that of $\angle X$. Find $\angle Z$. (Figure not drawn to scale.)

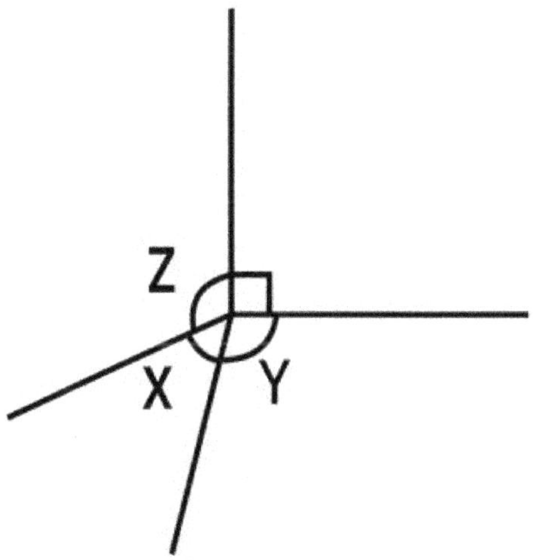

✓ Correct Ans **A.** 114°

B. 104°

C. 124°

D. 134°

Explanation

$\angle Y = 2 \times 52° = 104°$
$360° - 90° - 52° - 104° = 114°$
Hence $\angle Z$ is 114°

Q: 34 Cheryl earned a fixed income each month. In September, Cheryl spent some money and saved the rest. In October, the amount she spent increased by 20% and she saved 40% less compared to the previous month.

What fraction of Cheryl's income in September did she spend?

A. $^1/_3$

✓ Correct Ans **B.** $^2/_3$

C. $^4/_3$

D. $^3/_2$

Explanation

Total income remaindes same.
Sep : Income = p, spent = n, saved = p - n
Oct : Income = p, spent = nx1.2, saved = (p - n)x0.6

nx1.2 + px0.6 - nx0.6 = p
nx0.6 = px0.4
$^n/_p = {}^4/_6 = {}^2/_3$

Q: 35 In the figure shown below, QRUV is a square. Line QS is 15 cm and line WU is 35 cm. What is the perimeter of rectangle PSTW?

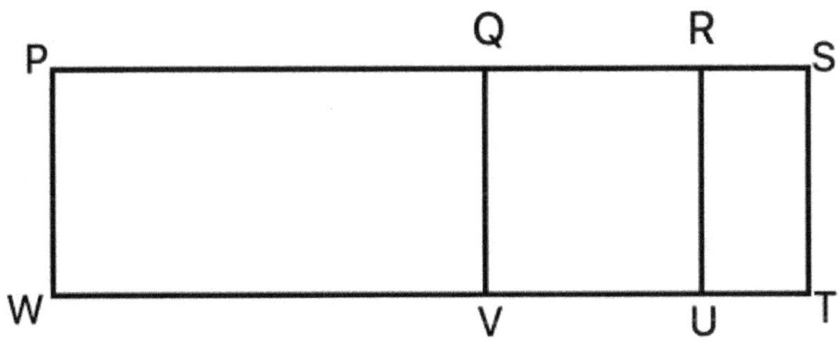

✓ Correct Ans **A.** 100 cm

B. 120 cm

C. 50 cm

D. 110 cm

Explanation

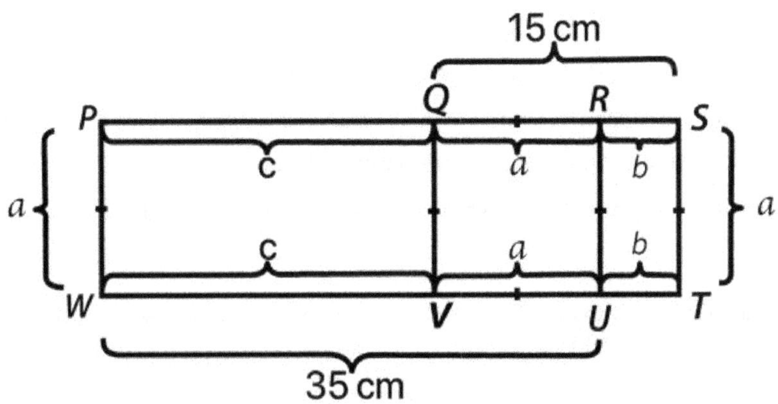

Perimeter → QR+RS+ST+TU+UV+VW+WP+PQ

= a + b + a + b + a + c + a + c = 15 + 15 + 35 + 35 = 100

The perimeter of rectangle PSTW is 100 cm

Q: 1

Dilan: "To have a greater chance of winning on Archery, you must practice for at least 30 minutes, four times a week."

Steve: "I do more than 30 minutes every day. I am so sure to win the Archery."

Which of these statements show the mistake that Steve has made?

A. The amount of practice time doesn't have any effect on the chance of winning.

✓ Correct Ans **B.** More practice time can not guarantee a greater chance of winning.

C. Less than 30 minutes of practice four times a week will lead to winning.

D. Less than 30 minutes of practice three times a week will lead to winning.

Explanation

Steve assumed more practice time than the minimum recommended can heighten the chance of winning on Archery, which is an incorrect assumption.

Q: 2 Mrs. Lee gives scores in lecture and laboratory to each student in his Chemistry class. These two scores are then added to give an overall score in Chemistry for the term (for example, a student with score 50 in lecture and 30 in laboratory would have an overall score of 80 in Chemistry.)

Last term Ace and Ben got the same overall score in Chemistry.

Ace: "If our scores in Lecture were the same, then our score in Laboratory must have been the same too."

Ben: "If our scores in Laboratory were different from each other, then our score in Lecture must have been different too."

If the information is true, whose reasoning is correct?

A. Ace only

B. Ben only

✓ Correct Ans **C.** Both Ace and Ben

D. Neither Ace nor Ben

Explanation

If the scores in Lecture were the same but the score in Laboratory is different then the total score could not be equal. This statement is true. And if the scores in Laboratory were different but the score in Lecture is same then the total score could not be equal. This statement is also true.

Q: 3 According to a survey of color preference: everyone who like pink liked red. Everyone who like red liked orange but no one who like red liked grey.
Kim, Loisa, Liza and Jane all took part in the survey.

Based on the information, which of the following must be true?

A. If Kim like orange, she also likes red.

B. If Loisa doesn't like grey, she doesn't like orange.

✓ Correct Ans **C.** If Jane likes pink, she does not like grey.

D. If Liza doesn't like pink, she doesn't like red.

Explanation

Construct a Venn diagram to arrive at the conclusion.

Q: 4 Kacey and her friends like fruits very much. Kacey loves oranges, bananas, and mangoes while Lily loves apples and oranges. Rachel doesn't like bananas and apples, but she loves all the other fruits. Which of the following is correct regarding the above situation?

A. Rachel likes all fruits that Kacey likes

B. Kacey doesn't like any fruit that Lily likes

C. Lily likes all fruits that Rachel likes

✓ Correct Ans **D.** Lily likes some fruits that Rachel likes

Explanation

Here are the likes of everyone.

- Kacey → oranges, bananas, and mangoes
- Lily → oranges, apples
- Rachel → all except apples and bananas

So, it is clear that Lily loves some fruits that Rachel likes.

Q: 5 Adam is a software developer. Usually, it takes 6 days to complete software development and if the client needs to get the work done in less than 6 days, the price will be inverse proportionally increased. e.g., if the client needs him to finish the software in 3 days, the price will be doubled.
If it usually costs $80 for a specific software development and a client needs Adam to finish it within 2 days, how much will it cost client?

A. 160

✓ Correct Ans **B.** 240

C. 40

D. 120

Explanation

The client needs the software within 2 days. That's $^1/_3$ of 6 days. So, the price will be 3 times higher. That is 80 times 3 = $240.

Q: 6 Jill has cricket practices **every** evening.
If she missed practices four days in a row last week, on which day did she definitely miss the practices?

Note : A week is Monday to Sunday

A. Saturday

B. Monday

✓ Correct Ans **C.** Thursday

D. Wednesday

Explanation

Possible ways :

- Monday, Tuesday, Wednesday, Thursday
- Tuesday, Wednesday, Thursday, Friday
- Wednesday, Thursday, Friday, Saturday
- Thursday, Friday, Saturday, Sunday

Thursday is in every possible way.

Q: 7 These are representations of three English words.

2465, 1465, 2769

What can the words be ?

A. Ring, Sing, Roll

B. Reel, Wheel, Real

C. Near, Node, Null

✓ Correct Ans **D.** Rank, Bank, Ring

Explanation

Look carefully at the repeated numbers and their positions. 1st and 3rd letters start with the same letter. 3rd letter of each word is the same.

Q: 8 Following tables show the marks of students on two science exams.

Exam 01

Hannah	60
Robin	55
Kathy	65
David	75

Exam 02

Robin	79
David	78
Hannah	82
Kathy	82

If the total marks of two papers of a student **who has made the least improvement** (from Exam 01 to Exam 02) is less than 135, he or she has to face a viva.

Who will have to face a viva?

A. Hannah

B. Kathy

C. David

✓ Correct Ans **D.** No one

Explanation

Improvements of each student.

- Hannah → 22
- Kathy → 17
- David → 6
- Robin → 24

David has the least improvement. His total marks are 75 + 78 = 153 which is greater than 135. So, no one has to face the viva.

Q: 9 Lily's cookery class teacher instructed to use : 2 eggs, 4 ½ cups of milk, 100 grams of sugar, and a teaspoon of salt to make the previously taught recipe for 2 people.
Lily wanted to make it this morning, and she needed 22 cups of milk. How many people are there at Lily's apartment?

A. 5

B. 6

C. 7

✓ Correct Ans **D.** 8

Explanation

Lily needed 22 cups of milk and that is four times 4 ½ cups of milk. That means she has used ingredients 4 times than for 2 people. So, there should be 8 people.

Q: 10 Barney went to buy a watch, and this is what he found out about the prices.
Piaget watches are expensive, but not like Rolex watches. Blancpain watches were more expensive than Longines watches, but not like Piaget ones. Omega watches are cheaper than Tudor watches, and are the cheapest of the lot.
Which brand has the 3rd cheapest watches?

A. Tudor
B. Rolex
C. Piaget
✓ Correct Ans **D.** Longines

Explanation

According to the prices of the brands, Rolex > Piaget > Blancpain > Longines > Tudor > Omega.
So, the 3rd cheapest one is Longines.

Q: 11 If it is Monday, George would attend cricket practices after school.
And also, if George doesn't attend practices, this means he is attending urgent tutoring class in the evening.

It is Monday and George has no tutoring class. So :

✓ Correct Ans **A.** He will attend practices
B. He won't attend practices
C. He will go home after the school
D. Cannot decide whether he attends the practices or not

Explanation

Given that it is Monday, and he has no classes, he will attend the practices after school.

Q: 12 **Weather is very important in farming. Every farmer knows well about the weather.**

Kevin: "My uncle knows well about the weather."

So...

A. Kevin's uncle is a farmer
B. Kevin's uncle is not a farmer
✓ Correct Ans **C.** Kevin's uncle may be a farmer
D. None of the above is correct

Explanation

If someone is a farmer, that person knows well about the weather. But there can be other people who know well about the weather, other than the farmers. However, C contains "maybe". So, C is correct because there is a possibility.

Q: 13 **Principal** : "The students of each class who got the highest marks for Mathematics will be given a prize by next week."

It's been two months since principal's announcement, but Daniel didn't get a prize. That means,

A. There was another student who got the lower marks

✓ Correct Ans **B.** Some students got higher marks than him in his class

C. Daniel misses everything

D. None of the above is correct

Explanation

If a student got the highest marks for Maths in his class, he will receive a prize. If someone does not receive a prize, then he did not get the highest marks.

Q: 14 Please refer to the statements below. For the purpose of this question, they are correct even if they contradict commonly known facts.

None of the vehicles are eco-friendly.
Some vehicles are electric vehicles.

Which of the following can be a conclusion ?

A. Only non-electric vehicles are not ecofriendly

B. Some vehicles are ecofriendly

C. Some electric vehicles are not ecofriendly

✓ Correct Ans **D.** All the electric vehicles are not ecofriendly

Explanation

If it is a vehicle, it is not eco-friendly. It can be either a car, bus, petroleum fuelled, or even electric. If it is a vehicle, it cannot be eco-friendly according to the first statement.

Q: 15 **Harry**: "If Emily goes to the party, she will be late to work tomorrow."

Rupert: "I think that won't happen."

What does Rupert think?

A. Emily won't go to the party

✓ Correct Ans **B.** Emily will go to the party, and she won't be late to work

C. Emily won't either go to the party or be late to work

D. None of the above

Explanation

Rupert thinks Harry is wrong. That means the negation of Harry's statement is true. That is, Emily goes to the party and she will not be late for work tomorrow.

Q: 16 Jenny and Joey were waiting at bus stop when they saw Emma was approaching the bus stop.

Jenney: "If it rains, Emma will miss the bus."

Joey: "Oh no. She gets angry whenever she misses the bus."

It is heavily raining since morning. Which is a next possible situation ?

✓ Correct Ans **A.** Emma is angry now

B. Maybe Emma is angry now

C. Emma is not angry

D. No conclusion

Explanation

According to Jenney and Joey, we can conclude that if it rains, then Emma gets angry. (hypothetical syllogism) Given that it's raining. So, Emma will get angry for sure.

Q: 17 Whenever Eunice is stressed, she takes a warm bath to relax. When she takes a warm bath, she eats her comfort food which is ice cream.

Teresa: "Eunice is stressed. She must be taking a warm bath to relax."

Jackie: "If Eunice took a warm bath then she must be eating an ice cream."

Whose reasoning is correct?

A. Teresa

B. Jackie

✓ Correct Ans **C.** Teresa and Jackie

D. Neither Teresa nor Jackie

Explanation

Stress -> Warm bath -> Ice cream. Both Teresa and Jackie said based on left-to-right reasoning, hence both are correct.

Q: 18 Carlo said that if Kim does not study hard during this semester then he will not pay for her tuition fee for the next semester.

Hanz: "If Kim studies hard on this semester then Carlo will pay for Kim's tuition fee next semester."

Betty: "If Carlo did not pay for Kim's tuition fee for this semester then Kim must've not studied hard last semester."

Whose reasoning is correct?

A. Hanz

B. Betty

C. Hanz and Betty

✓ Correct Ans **D.** Neither Hanz nor Betty

Explanation

Left to right reasoning : No study -> No payment for next semester.
There is no gurantee that study would result in payment for next semester.
Also Betty is talking right to left, hence not guranteed too.

Q: 19 When Emily told her friend Alex that she was considering skipping the morning run to focus on her work, Alex said: "You don't have to skip the run entirely. Regular exercise, such as running, helps improve focus and productivity."

Which of the following statement **strengthens** Alex's argument?

A. Emily's work performance has been consistently poor.

B. Alex is a professional athlete who manages both running and work responsibilities.

C. Running is not effective for improving focus and productivity.

✓ Correct Ans **D.** Many successful professionals attribute their efficiency to regular exercise routines.

Explanation

Option D strengthens Alex's argument by stating that many successful professionals credit their efficiency to regular exercise routines. This supports Alex's claim that regular exercise like running can enhance Emily's focus and productivity.

Q: 20 A language school offers an advanced level course for students who have completed the intermediate level or can hold conversations fluently. Students who excel in grammar and writing are invited to a special writing workshop. Among the students Jake, Emily, Liam, and Olivia participated.

Which conclusion is supported by the information?

A. If Jake is invited to the writing workshop, he has completed the intermediate level.

✓ Correct Ans **B.** If Emily can hold conversations fluently, she is eligible for the advanced level course.

C. If Liam completed the intermediate level, he is invited to the writing workshop.

D. If Olivia is not invited to the writing workshop, she is in the intermediate level.

Explanation

The information suggests that fluent conversation skills are a requirement for the advanced level course. Therefore, if Emily can hold conversations fluently, she is indeed eligible for the advanced level course.

Q: 21 A kind of vitamin B3 is nicotinamide, often known as niacinamide. In addition to meat, fish, milk, eggs, green vegetables, and cereals, it may also be found in many other foods. For carbohydrates and fats to operate properly in the body and to keep cells healthy, niacinamide is necessary. A key component of skin whitening, niacinamide prevents the transfer of melanin pigment to skin cells. After being used for a while, niacinamide reduces hyperpigmentation and progressively lightens skin.

If the information is true, which of the following is **out of context** ?

A. Niacinamide progressively lighten the skin and reduces hyperpigmentation.

✓ Correct Ans **B.** Niacinamide can be taken to genetically change any dark skin color.

C. Niacinamide can be found on fish, milk, eggs, green vegetables, cereals and meat.

D. Niacinamide is a kind of vitamin B3.

Explanation

Concept of genetics is out of context.

Q: 22 In a certain language, the word UNCLE is written as WPENG. How will you write the name AUNT?

A. BVOU

✓ Correct Ans **B.** CWPV

C. DXQW

D. None of the above

Explanation

U is represented by W, which is 2 steps ahead; P represents N, and it is also the letter after N and so on. Therefore, C represents A, U ->W, P->T and T->V. The word should be CWPV.

Q: 23 Which one of the following words comes last in the dictionary?

A. Achieve

B. Acquit

C. Accumulate

✓ Correct Ans **D.** Actuate

Explanation

Whereas the first and the second letters are common, Ac, the third ones aren't, and they are h, q, c, and t. Since t is the last alphabet letter among the 4, Actuate will come last in the dictionary.

Q: 24 **Which answer suits the given pattern?**

ABDG, IJLO, QRTW, ...

1. EFHJ

✓ Correct Ans **2.** GHJM

3. NPSW

4. KMOR

Explanation

Each word has four letters according to the following pattern.
Any letter, the next letter, the letter one after the previous letter, the letter two letters after the previous letter.

Q: 25 ***Emily ate a lot of sweets.***
If Emily sleeps without brushing her teeth, she will likely to get toothache tomorrow.
If Emily got a toothache, she will go to the dentist.
If she goes to the dentist, she won't be able to participate in the lectures.

If the above information is true which of the following is impossible?

A. Emily went to the dentist but she did not brush her teeth before sleeping.

B. Emily didn't visit the dentist but she didn't attend the lectures too.

C. Emily brushed her teeth before sleep, but she went to the dentist.

✓ Correct Ans **D.** Emily got a toothache but she participated in the lectures too.

Explanation

If Emily got a toothache she will go to the dentist and won't be able to participate in the lectures. So statement D is impossible.

Q: 26 There was a banner near the road displaying "**Speed thrills but kills.**"

Which one of the following statements supports the above slogan the best?

A. Driving at high speed is an action-packed experience

✓ Correct Ans **B.** We should drive at a normal pace as high speed cause fatal accidents and reduce the chance of avoiding an accident, It may cost not only our life but even others

C. Watching high-speed vehicles are very thrilling, but waiting near a road is not safe as road accidents can be fatal.

D. Though driving at a high speed is a very thrilling experience, it's harmful and damages your vehicle a lot.

Explanation

The statement should provide a reason to support the slogan that speed might be thrilling but it might kill you too. Option B supports the slogan best.

Q: 27 A paediatrician told: if your kid keeps a soft toy with her, don't forget to clean it daily. If your kid is suffering from asthma or other respiratory problems, then you should keep those stuffed toys away from your child. Otherwise, the teddy bear also can be harmful as a real bear.

Which of the following, if true, most strengthens the paediatrician's claim?

A. When a child loves teddy bears she might tend to think that real bears are not harmful too, which is not so good.

B. Soft toys quickly get dirty so it's essential to clean them daily otherwise the child will get sad and depressed.

✓ Correct Ans **C.** Soft toys are home to dust mites and other allergens which are common triggers for allergy and other respiratory symptoms.

D. Soft toys and plush animals will help them reduce these feelings and stressful emotions.

Explanation

The statement should provide a reason to support the claim that teddy bears are harmful to the child's health, especially for the respiratory system. This is option C.

Q: 28 Robbie started reading today's newspaper. There is an advertisement in first page that said :

"Buy pure and natural honey of company Golden Honey"

There are assuptions behind the above statement from advertisement.

Assumption I: No other company supplies pure and natural honey.
Assumption II: People read the advertisements.

Which assumption(s) is/are implicit ?

A. Only assumption I is implicit.

✓ Correct Ans **B.** Only assumption II is implicit.

C. Neither assumption I nor II is implicit.

D. Both the assumptions are implicit.

Explanation

Advertisements cost money for advertiser. A company is spending on advertisement because the company believes that people will see the advertisement and respond positively.

Q: 29 All the sunflower in Ethan's garden is White.
All the Roses in Ethan's garden are Pink.
All the flowers in Ethan's garden are either White or Pink.

If the first two statements are true, the third statement is...

A. True

B. False

✓ Correct Ans **C.** Uncertain

D. Can't decide

Explanation

We don't know what else flower plants Ethan has in his garden.

Q: 30 There are 53 students in a class. 20 students have more than 5 pens each and 15 students have 4 pens each and the remaining students have one pen each.

Melinda: "35 students have more than 2 pens each"
Harry: "The number of students not having more than 5 pens are 15"

Whose argument is correct?

✓ Correct Ans **A.** Melinda is correct

B. Harry is correct

C. Both Melinda and Harry is correct

D. Neither Melinda nor Harry is correct

Explanation

Students having more than 2 pens each = 20 + 15 = 35. Hence, Melinda is correct.

Q: 31 EVS, English and Maths books are kept in the manner explained.
Maths book is not the last book.
EVS book is above the maths book .
English book is not at second place.

Which book is the first book from the top?

✓ Correct Ans **A.** EVS

B. Maths

C. English

D. Can't be determined.

Explanation

Construct a diagram with vertical line to arrive at the answer.

Q: 32 Consider the following conditions regarding Samantha's marks.

If Samantha's marks are greater than 75, she will receive a gift.
If Samantha's marks are greater than 60, she can celebrate it.
If Samantha's marks are less than 65, she has to study well.
If Samantha's marks are less than 62, 2 penalty marks will be reduced.

Samantha got 62 marks for the exam.

Select the most suitable outcome.

1. Samantha can celebrate, but she will receive 2 penalty marks.
2. Samantha's new marks are 60.
3. Samantha will receive a gift, but she has to study well.
4. ✓ Correct Ans — Samantha can celebrate, but she has to study well.

Explanation

Samantha's marks are less than 65, greater than 60, and still not less than 62. Marks will only be penalized if it is less than 62.

Q: 33 Handfish are a type of anglerfish famous for their **very unique** modified fins that resemble human hands, which they use to "walk" across the sea floor rather than swim.

Mandy: "If you see a fish with modified fins that resemble human hands it must be a Handfish"

Kara: "Also if you see a creature walk across the sea floor rather than swim, it must be a Handfish"

If the information given is correct, whose reasoning must be correct?

- ✓ Correct Ans — **A.** Mandy only
- **B.** Kara only
- **C.** Both Mandy and Kara
- **D.** Neither Mandy and Kara

Explanation

Modified fins that resemble human hands are a unique feature of handfish. So Mandy is correct.
But there might be many other creatures who walk across the sea floor rather than swim. So Kara is wrong.

Q: 34 **Find the first two terms of the following sequence.**

A, B, 24, 26, 52, 54, 108, ...

1. 10, 20
2. 6, 12
3. 10, 12 ✓ Correct Ans
4. 6, 8

Explanation

The sequence goes like adding two to the previous term then doubling the term. After the first two terms, 24 has become 26 which means 24 + 2. Then 26 has been doubled which is 52.
So,B should be the half of 24. so, it is 12. A should be two less than be. then it is 10.

Q: 35 Read the following statements carefully.

- If someone goes to the gym, then he becomes healthy.
- Healthy people are attractive.
- If someone is attractive, then he will have a lot of friends.

My friend Kevin goes to the gym.

Which answer is the **most correct** about Kevin?

1. Kevin is not attractive.
2. Kevin is healthy but he doesn't have a lot of friends.
3. Kevin is healthy and attractive, and he has a lot of friends. ✓ Correct Ans
4. Kevin's friends are attractive.

Explanation

Kevin goes to the gym so he's healthy and attractive and hence he has a lot of friends.

Q: 36 I have 10 friends. Five of them are vegetarians and two of them don't like ice cream.
Six of my friends have pet dogs and others have cats.

Which of the following is NOT correct about my friends?

1. Some of my friends have pet dogs.
2. If someone is a friend of mine, he likes ice cream. ✓ Correct Ans
3. I have some friends who like ice creams.
4. All of my friends have dogs or cats.

Explanation

Given that two of my friends don't like ice cream, means are there 8 friends who like ice cream. So, only some of my friends like ice cream. Therefore we cannot say that if someone is a friend of mine he likes ice cream.

Q: 37 **Consider the following statements.**

Tasty fruits are fragrant and contain vitamin C.
James found a fruit from a forest and it's fragrant.

Select the correct conclusion from the above statements.

1. If it contains vitamin C, it is tasty.
2. The fruit is not tasty.
3. ✓ Correct Ans — If it does not contain vitamin C, it is not tasty.
4. It is tasty because it is fragrant.

Explanation

According to the first statement if the fruit is tasty then it is fragrant and contains vitamin C. From the contra positive of this statement, if the fruit does not contain vitamin C or it is not fragrant then it's not tasty. James found that the fruit is fragrant. So if it does not contain vitamin C, it's clear that the fruit is not tasty from the contrapositive of the statement.

Q: 38 If 2-3-5-8-1-6 is GROUND and 8-9-0-0 is BALL, what is the meaning of 8-3-4-2-9-6-7?

A. ✓ Correct Ans — BRIGADE
B. BRING
C. BRIGADIER
D. BRIGAND

Explanation

Representations of letters :
G→2, R→3, O→5 U→8, N→1, D→6, B→8, A→9, and L→0

So, 8-3-4-2-9-6-7 is BR_GAD_. The only answer that suits this result is BRIGADE.

Q: 39 Antioxidants and minerals like calcium and potassium are abundant in radishes. These nutrients work as a team to lower high blood pressure and lower your risk of developing heart disease. The radish is a good source of naturally occurring nitrates that enhance blood circulation.

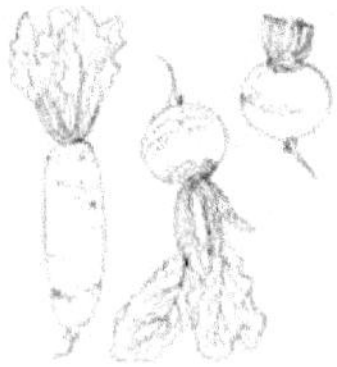

If the information is true, which of the following must be false?

A. Radish enhances blood circulation.

B. Nutrients in radish lowers high blood pressure.

C. Radish are abundant in calcium and potassium.

✓ Correct Ans **D.** The antioxidants and minerals in radish increases the risk of developing a heart disease.

Explanation

Antioxidants and minerals like calcium and potassium are abundant in radishes. These nutrients work as a team to lower high blood pressure and lower your risk of developing heart disease. So, the antioxidants and minerals in radish increases the risk of developing a heart disease is false.

Q: 40

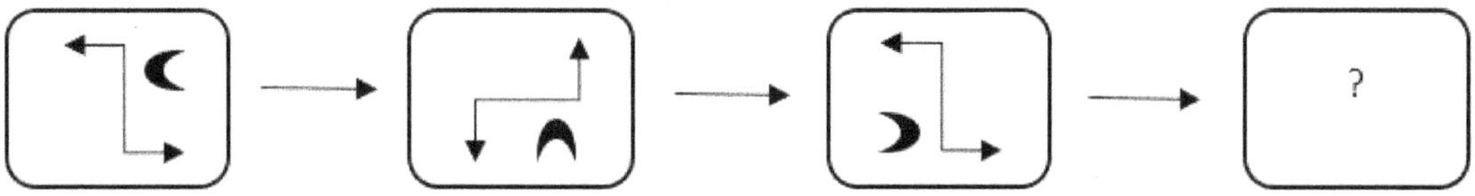

✓ Correct Ans **A.**

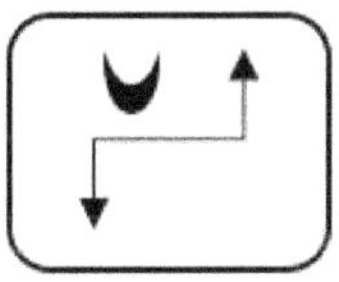

B.

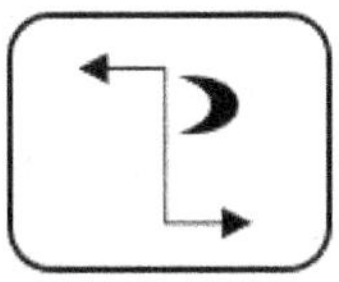

C.

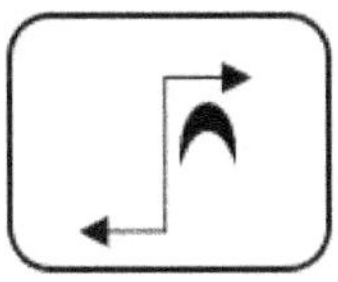

D.

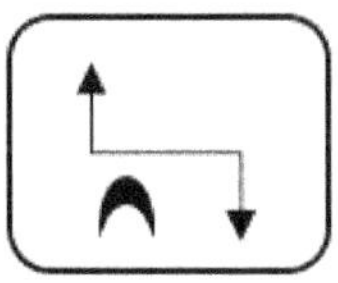

Explanation

It can be observed that in each figure transition, the arrow is rotating clockwise in 90 degrees and the moon is moving along with the arrow. Thus, the last figure is option A.

Printed in the USA
CPSIA information can be obtained
at www.ICGtesting.com
CBHW080744270824
13727CB00049B/810

9 781922 608345